Decentralisation and Empowerment

The Odisha Model for Urban Development

G. Mathi Vathanan
Shubhagato Dasgupta

IWA PUBLISHING

Published by **IWA Publishing**
Unit 104–105, Export Building
1 Clove Crescent
London E14 2BA, UK
Telephone: +44 (0)20 7654 5500
Fax: +44 (0)20 7654 5555
Email: publications@iwap.co.uk
Web: www.iwaponline.com

First published 2025

British Library Cataloguing in Publication Data
A CIP catalogue record for this book is available from the British Library

ISBN: 9781789064919 (paperback)
ISBN: 9781789064926 (eBook)
ISBN: 9781789064933 (ePub)

Doi: 10.2166/9781789064926

This eBook was made Open Access in January 2025.

Contents

Reviews of the Book

Ms Anacláudia Rossbach
Under-Secretary-General and Executive Director of the United Nations Human Settlements Programme (UN-Habitat)

Odisha's development journey holds remarkable learnings for cities embarking on a path towards inclusive urbanisation around the world. The state has spearheaded a forward-looking agenda by adopting decentralisation in its true spirit by devolving funds, functions and functionaries to local levels. Amongst the several noteworthy aspects of Odisha's transformative programmes are the focus on strengthening legislations, participatory and transparent community engagement, and environmental and financial sustainability as key enabling factors for impact merit special attention. Each of these aspects strongly resonates with UN-Habitat's emphasis on the 'People's Process' and aligns with the development pathways under the New Urban Agenda.

Odisha paves the way to successfully anchor guiding principles, 'decentralised, inclusive and community led', in complex urban domains of land rights for slum dwellers, water supply for all, innovations in sanitation and septage management, solid waste management, and food as a right, among others. By empowering local communities, women's self-help groups and slum dwellers' associations in co-creation and co-implementation of projects, the state has fostered a sense of shared partnership between local communities and local governments.

Dr Asit K. Biswas
Academician and Distinguished Visiting Professor, University of Glasgow, UK
Chief Executive, Third World Centre for Water Management, Mexico
Stockholm Water Prize Laureate, 2006

As one who grew up in Baleswar, Odisha, this book by Mathi Vathanan and Shubhagato Dasgupta is of great personal interested to me. For decades, I have been advocating that, while there are many similarities in provisioning urban water supply all over the developing world, the solutions have to be local and context specific. Recently, Odisha has witnessed a rapid transformation in provision of urban water and sanitation services. This remarkable advance within a short time frame was possible primarily through decentralisation of authority and empowerment of communities through participation in service deliveries. These remarkable achievements, including drink from tap 24/7 water supply and city-wide inclusive sanitation, are excellent examples of what can be achieved in India and elsewhere through good planning and execution.

The book provides in-depth analyses of enabling conditions that made such achievements possible. Provision of water supply and sanitation is not rocket science. We have known for decades what needs to be done, but, for various reasons, this has not been done in Indian urban centres.

I strongly recommend this book to policymakers, water utility managers, academics and students, to learn how and why Odisha became a success story whereas many other similar attempts have failed.

Dr Amita Bhide
Professor, Centre for Urban Policy and Governance, School of Habitat Studies, Tata Institute of Social Sciences, Mumbai, India

Odisha's urban transformation is perhaps one of the most inspiring stories of change seen in India in recent times. In a context where urbanisation either stands for elite oriented exclusionary cities spurred by a thirst for international investment, or unplanned, poorly serviced and haphazard growth of settlements with highly debilitating quality of life characterised by significant spatial, and socio-economic inequities, the trajectory that Odisha has chosen is one that stands for its distinctive path. This path has prioritised universalisation of services, appropriate use of technology and resources over high-end technology, invested in people and particularly poor people and built their institutions, built partnerships with a range of stakeholders and balanced welfare with reforms. What is even more astounding is the fact that less than two decades ago the state was conventionally known as a 'laggard' state in terms of urban governance and infrastructure.

This book documents this comprehensive process of change authored by two key agents of this urban transformation that is strategic, and has demanded a significant political will and shift in policy directions. It gives some insight into the nature of this process which is highly relevant to the cities of the Global South. Therefore, it is a highly valuable book for urban planners, policy makers, researchers and practitioners.

Mr Barjor Mehta
President, CEPT University, Ahmedabad, India

This book is a valuable addition to literature that will benefit students and practitioners of urban development and planning, both in India and in other countries facing similar challenges. It documents urban development programmes that have successfully expanded basic services across all cities in

Odisha State over the past decade. A key highlight of the book is its coverage of lessons learned across an impressive range of municipal sectors, while also offering general approaches to the delivery of city services. Odisha's progress in decentralisation and empowerment over the last decade stands out as a shining example of urban development in India.

The authors of this book were actively involved in designing and implementing the innovative urban programmes it presents, allowing them to share unique insights and lessons that emphasise the value of decentralisation for service delivery, as well as participatory and community-driven planning. Notably, the book explores how such processes can be structured to operate efficiently and achieve impactful results at scale within a relatively short period.

This book is highly recommended for city leaders, specialists, and students alike – and is a must-read for anyone interested in urban issues.

About the Book

Decentralisation and Empowerment: The Odisha Model for Urban Development discusses the key approaches and factors that underlie the highly successful urban development initiatives implemented in Odisha for close to a decade. The application of the governance principles of decentralisation and empowerment in a consistent, contextual, strategic and tactical manner across sub-sectors, be it drinking water supply, environmental sanitation, municipal waste management, slum land rights, infrastructure upgrading or others, has led to the development of the Odisha Model. The book details the conceptualisation, design, learning and implementation processes that underpinned achieving, with remarkable speed and scale, the various urban initiatives rolled out in the State. Most of these interventions have been accorded multiple national and global accolades.

The book documents and deliberates on the reasoning behind the various innovative approaches designed and adopted while planning, implementing and monitoring these initiatives. It provides an insider's perspective on the thought processes that informed the development of public programmes that focus on the wellbeing of citizens. Therefore, this book will be of immense interest and use to the academic and research community including students. It is a source of inspiration for what can be achieved, and a valuable guide for other governments, city administrations, specific sector professionals and urban practitioners.

The book has been co-authored by Mr Mathi Vathanan, a senior Indian Administrative Service officer of the Odisha cadre (presently Director General of the State Administrative Training Institute) who headed the Housing and Urban Development Department under whose leadership the documented initiatives were crafted, and by Mr Shubhagato Dasgupta, a well-regarded urban development professional, Senior Fellow at the Centre for Policy Research and presently also Team Lead, SCIAI programme at GWSC, AIT, Thailand, who have worked closely together since 2015.

About the Authors

G. MATHI VATHANAN, IAS
Director General, State Administrative Institute and Head of Centre for Good Governance, Odisha
Email: egmv66@gmail.com

Mr G. Mathi Vathanan is a senior Indian Administrative Service Officer of Odisha cadre, presently posted as the Director of the State Administrative Training Institute and Head of Centre of Good Governance, Odisha. Until July 2024, he was the Additional Chief Secretary of the Department of Housing and Urban Development Government of Odisha. He led a variety of initiatives, including the Jaga Mission, which provides slum dwellers with land rights and upgrades, and the 24 × 7 Drink from Tap Water Supply programme, which offers every urban household direct access to clean drinking water. His approach of a decentralised strategy with community partnership has become a central aspect of the Odisha model of urban governance. These initiatives have earned recognition, such as the World Habitat Awards in 2019 and UN Habitat: World Habitat awards in 2023.

Under his guidance, Odisha has achieved 100% water supply house connections in 105 cities out of 115, and Puri and Gopalpur have become the only two towns in India to achieve a 24 × 7 Drink Directly from Tap Quality water supply system. Furthermore, he is promoting city-wide inclusive sanitation in all 115 cities in Odisha. This work has been recognised globally, including the Global Water Leader Distinction Award given to WATCO at the Global Water Summit in Spain in 2022 and WRI: Ross Prize for cities Award, 2022 for the Urban Wage Employment scheme.

Since 2023, he has been a member of the Advisory Board for IWA's Inclusive Urban Sanitation initiative. His previous book, *People First: How Odisha's Drink from Tap Mission Quenched Every Thirst*, was published in 2024 by Rupa Publications.

SHUBHAGATO DASGUPTA
Senior Fellow, Centre for Policy Research; Director, Scaling City Institutions for India (SCI-FI), New Delhi, and Team Leader, SCIAI@GWSC, Asian Institute of Technology, Thailand.
Email: shubhagato@hotmail.com

Shubhagato Dasgupta is a well-regarded urban professional and researcher with close to 30 years of experience in India and elsewhere in South Asia, East Asia and Africa. He has worked with public, private, multilateral, non-governmental, and academic organisations that have contributed to his unique, holistic understanding of urban development and housing issues and solutions.

In the public sector, when he was at the Housing and Urban Development Corporation (HUDCO), he was involved in drafting national policies, supporting the setting up of retail housing finance, organisational restructuring and project evaluations. Later, he led the Support to National Policies for Urban Poverty Reduction project, a collaboration between the UK's Department for International Development (DFID) and India's Ministry of Housing and Urban Poverty Alleviation. This project focused on developing pro-poor urban and housing policies (including RAY, PMAY, NULM, and the Credit Guarantee programme) in 20 cities across 15 states of India.

In the private sector Shubhagato worked with the Infrastructure Development Finance Corporation and others, where he led private investments in urban infrastructure projects and structured a number of flagship urban infrastructure and housing PPPs. He has also worked with multilaterals as Senior Urban Specialist at the World Bank and International Finance Corporation, helping design public programmes like the City Challenge Fund and JNNURM in India, and small-scale public-private infrastructure programmes in Africa. He has also worked on innovative programs with NGOs (TARU and ASAG) and policy research programs with think tanks (CPR-India and ISAS-Singapore) and academic and training organisations (HSMI and GWSC-AIT).

He has the rare experience of being closely involved in designing large, pioneering and transformative public and private programmes at national, state and city levels. He has successfully built and delivered large-scale research and practice programmes (Scaling City Institutions for India), which involved research to develop transformative policies and programmes, including some of the initiatives in Odisha discussed in this book.

Shubhagato trained as an architect at the Centre for Environmental Planning and Technology in Ahmedabad and holds an MSc in housing and development planning from the Development Planning Unit at University College London. He is currently undertaking a PhD at the University of Amsterdam.

His present research interests are focused on urban resilience issues, drinking water, sanitation, waste management and housing. His contribution to service delivery challenges in smaller cities and large dense villages is highly regarded. He has authored several journal articles, and his book *Social Innovations in Urban Sanitation in India: Meeting Unmet Needs* (along with his co-authors) was published by Routledge in 2023.

Acknowledgements

The authors would first like to extend their immense gratitude to the Government of Odisha (Team Odisha) and the people of Odisha without whom the Urban Odisha Model and this book would not have been possible.

The authors are grateful to the members of the Urban Odisha Team – Programme Heads and their team members, ULB Teams including the community partners (SHGs, SDAs, etc.), and specifically Sangramjit Nayak, Kalyan Rath, Sushant Mishra, Sagarika Patnaik, Sarada Panda, Durgesh Nandini, Er. P.K. Swain, Er. P.K. Mohapatra, Shishir Dash, Chinmay Tripathy, Anindya Panda, Pranati Dash, Rashmita Patel, Srinivas Alavalli, Suchisnata Sahoo, Subhranshu Satpathy, Antarin Chakraborthy and Pinakapani Parida.

The authors would like to acknowledge the support of researchers for various aspects of the book, including Baisakhi Sarkar Dhar, Shaurya Gupta and Anindita Mukherjee, for their support throughout the process of writing and publishing this book, Arushi Gupta for chapters 2, 3 and 13; Baisakhi Sarkar Dhar for chapters 8, 9 and 12; Neha Agarwal for chapter 5; Shikha Shukla Chhabra for chapters 6 and 7; Tanvi Tomar for chapter 4; Tripti Singh for chapter 10 and Abhinav Kumar for chapter 11.

The authors want to thank Nikhil Pawan Kalyan, Sudharshan Chakravarthy, Chakravarthi Singh Rathore, Atanu Kumar Samanta, Dilip Kumar Mohanty, Harsh Kothari and Biraja Satpathy for sharing their views on the implementation of various programmes.

We would like to thank our institutional partners who have played important roles in various initiatives of the government from conceptualizing to implementation support, including the Bill and Melinda Gates Foundation, Omidyar Network, Tata Trusts, Tata Steel Foundation, Centre for Policy Research, UNICEF, e-Gov Foundation, Janaagraha, Urban Management Center, Centre for Science and Environment, NIUA and others.

The authors would also like to thank the IWA Publishing team for their meticulous help in bringing out this book, especially Ms Katharine Allenby and Mr Andrew Peart.

Finally, we are truly grateful to the kind reviewers who have provided encouraging words about this book: Ms Anacláudia Rossbach, United Nations Under-Secretary-General and Executive Director, UN-Habitat, Nairobi; Dr Asit K. Biswas, Distinguished Visiting Professor, University of Glasgow, Chief Executive, Third World Centre for Water Management, and Stockholm Water Prize Laureate, 2006, Glasgow; Dr Amita Bhide, Professor, Centre for Urban Policy and Governance, School of Habitat Studies, Tata Institute of Social Sciences, Mumbai; and Mr Barjor Mehta, President, CEPT University, Ahmedabad.

List of Abbreviations

AMRUT	Atal Mission for Rejuvenation and Urban Transformation
ALF	Area Level Federation
BMGF	Bill & Melinda Gates Foundation
CLF	City Level Federation
CPHEEO	Central Public Health and Environmental Engineering Organisation
CPR	Centre for Policy Research
CSR	Corporate Social Responsibility
CTs/PTs	Community Toilets/Public Toilets
DFT	Drink from Tap
DMA	Directorate of Municipal Administration
DTP	Directorate of Town Planning
EO	Executive Officer
EWS	Economically Weaker Section
FCDO	Foreign Commonwealth & Development Office
FSSM	Faecal Sludge and Septage Management
FSTP	Faecal Sludge Treatment Plant
GOI	Government of India
HUDD	Department of Housing and Urban Development
IA	Implementing Agency
IHHL	Individual Household Latrine
JNNURM	Jawaharlal Nehru National Urban Renewal Mission
LIG	Low-Income Group
MC	Municipal Commissioner
MoUD	Ministry of Urban Development
MSG	Mission Shakti Group
NIUA	National Institute of Urban Affairs
NSSO	National Sample Survey Office
NUA	New Urban Agenda
ODF	Open Defecation Free
OLHM	Odisha Liveable Habitat Mission
OSCSK	Odisha State Commission for Safai Karamchari

OSPCB	Odisha State Pollution Control Board
OSS	On-Site Sanitation
OUHM	Odisha Urban Housing Mission
OUSM	Odisha Urban Sanitation Management
OUSP	Odisha Urban Sanitation Policy
OUSS	Odisha Urban Sanitation Strategy
OWSSB	Odisha Water Supply and Sewerage Board
PHEO	Public Health Engineering Organisation
PMAY	Pradhan Mantri Awas Yojana
PPP	Public Private Partnership
SDA	Slum Dwellers' Association
SDGs	Sustainable Development Goals
SHG	Self-Help Group
SLNA	State Level Nodal Agency
SOP	Standard Operating Procedure
SSB	Social Security Benefits
STP	Sewage Treatment Plant
SWM	Solid Waste Management
UASRRC	Urban Area Slum Rehabilitation and Redevelopment Committee
UGI	Urban Governance Index
ULB	Urban Local Body
UMC	Urban Management Centre
UNICEF	United Nation's International Children's Fund
USAID	United States Agency for International Development
WATCO	Water Corporation of Odisha

doi: 10.2166/9781789064926_0001

Chapter 1
Introduction

According to the 2011 Census of India, the urbanisation level in Odisha was at 31.16%, securing for the state the 18th rank among India's 28 states in terms of urbanisation. While Odisha is one of the least urbanised states in India, it registered remarkable urban growth during 2001–2011. The urbanisation pattern in Odisha has been characterised by the growth of both small and large cities, as well as the development of new planned cities.

In the present pattern of Odisha's urbanisation, smaller cities and towns have mushroomed across the state over the last decades, while the larger cities are fewer. Urban areas have been offering citizens expanded gamut of opportunities. The urbanisation trend over the last five decades in Odisha reveals the fast-paced growth of smaller towns in comparison to their larger counterparts.

In India, the push to decentralise urban governance and devolve powers came in the form of the 74th Amendment to the Constitution in 1992. This landmark constitutional intervention has formally given urban local bodies (ULBs) more autonomy and financial powers and recognised them as the third tier of government (with the central and state governments as first and second tiers, respectively). The Amendment has ushered an era of democratic decentralisation incorporating specific provisions for the empowerment of local governments (Mathur, 2013). The 74th Amendment lists under Schedule 12 several functions that are to be the responsibility of local bodies: such as urban planning, including town planning; regulation of land use; construction of buildings, roads and bridges; and the provision of water, public health, sanitation and solid waste management.

In Odisha, urban areas are governed by different tiers of local bodies, including Municipal Corporations, Municipalities and Notified Area Councils

(NACs). A Municipal Corporation, also known as a Nagar Nigam, is the highest tier of urban local government in the state. It governs large cities with a population of over 100 000 and is responsible for providing a range of services to the citizens, including water supply, sanitation, health and hygiene, and solid waste management. Currently, there are five Municipal Corporations in Odisha: Bhubaneswar, Cuttack, Berhampur, Sambalpur and Rourkela.

A Municipality, also known as a Nagar Panchayat, governs smaller urban areas with a population of 20 000–100 000. It is responsible for providing similar services to the citizens as a Municipal Corporation. An NAC is the lowest tier of urban local government in the state and governs small urban areas with a population of less than 20 000. It is responsible for providing basic services to the citizens, such as water supply and sanitation.

Each tier of local government in Odisha has its own elected representatives, who are responsible for governing the respective urban area and making decisions about the provision of services and infrastructure. The state government provides financial and administrative support to these local bodies to help them carry out their responsibilities effectively.

Numerous scholars have contended that the incomplete implementation of the 74th Constitutional amendment in India has manifested some common problems such as a financial gap and lack of spatial planning faced by the majority of cities, linked to their weak institutional capacities and sub-optimal capacity of their administration (Shastri & Singh, 2023; Sharma, 2014; Bercegol, 2017). As discussed by scholars given the current context of higher capacities in larger cities compared to smaller cities the impact of the 74th Constitutional Amendment on intra-urban disparity could be adverse, leading to further socio-economic segmentation (Kundu, 2007). It is in this context that the small town-led urban development of Odisha calls for an in-depth understanding into the state's management of urban development.

The urban expansion of Odisha and the growth of small towns has placed an enormous strain on land, housing supply and infrastructural services. *Decentralisation and Empowerment: The Odisha model of urban development* is an extensive documentation of the key urban development initiatives of Odisha, the context and the thought processes that have been driving them, as well details of the strategic and policy actions accompanying this that have received numerous national and global accolades. It covers key urban governance- and management-related interventions undertaken by the Government of Odisha, since 2013/2014.

The documentation also aims to present various innovative approaches adopted for planning, implementing and monitoring these initiatives. For instance, the Government of Odisha has chosen a strategic approach – through land management reforms and urban planning – for resolving land issues. Through exclusive insights from Odisha's model of urban development, the book comprehensively details some of the best practices of urban governance in the country that directly contribute to the overall wellbeing of the citizens. It argues for the need of optimal urban management, especially in smaller cities, with evidence from Odisha's journey and internationally acknowledged success in urban management.

This book will add to the limited literature on the internal decision-making processes of urban development management in India. It is hoped that this book will offer deep insights for other state and national governments; urban development experts and practitioners; academicians and researchers; student and urban historians and citizens at large into how and why urban development policies and programmes take particular form. It is also envisaged that this book will be of immense value to state administrators, urban planners and city managers seeking to address various challenges in the urban sector, especially for the most disadvantaged population groups in an urban setting. It is also hoped that the book will encourage other geographies to undertake urban development reform and encourage wider application of similar innovative models of urban development.

As the chapters in the book will outline, the Government of Odisha has introduced multiple urban development initiatives over the last few years, emerging as a pioneer in implementing successful, welfare-oriented urban interventions.

This chapter seeks to lay out the context of urban Odisha for the readers, to help them attain a better understanding of the urban development schemes, programmes and policies discussed in the subsequent chapters.

Chapter 2 on *Urbanisation and Economic Growth* lays out Odisha's situation as one of the lower urbanised states in the country, estimating that Odisha's urbanisation is poised to accelerate in the coming years, with a gradual decline in the state's share of rural population. Chapter 2 unpacks the trajectory of Odisha's urbanisation, along with the demographic, economic and social aspects that have shaped it. Some of the key points discussed in Chapter 2 include the state's population growth trends, urban growth pattern, namely, spatial distribution of towns, urban inequalities and slum formation, and economic indicators such as gross state domestic product (GSDP), per capita income, workforce participation rate and so on. It also highlights some human development indicators such as sex ratios, status of education and public health vis-à-vis gender and caste.

Chapter 3 on *Odisha's Devolved, Innovative Urban Governance* outlines the urban development institutional structure in the state. It discusses the vision, mission, functions and objectives of state's nodal unit for planned urban growth, the Housing and Urban Development Department (HUDD), and its directorates (attached offices). The purpose of Chapter 3 is to map the governance framework for urban Odisha, establish the functionalities of the agencies under the aegis of the HUDD and outline the hierarchical structure of the officers, to better describe the roles and duties of each actor at the state/district/ULB level in the implementation of the programmes discussed in the book.

Chapter 4 on *State-Wide Environmental Sanitation through Innovations in Faecal Sludge & Septage Management* focuses on the innovative and pioneering policy interventions and projects undertaken by the Government of Odisha in sanitation that has become a national and global good practice model. The state adopted a faecal sludge and septage management (FSSM) approach for ensuring environmentally sustainable sanitation at a time when achieving improved urban sanitation was understood as implementing underground sewerage systems based on concepts from the 1850s. It offers a brief analysis of the existing sanitation situation and the treatment capacity of the state along with tracing the state's sanitation journey in both urban and rural areas. Chapter 4 deals with an in-depth review of policy interventions (Odisha Urban Sanitation Policy, Odisha Urban Sanitation Strategy, other regulations, etc.) made to create an enabling environment at the state level. It then details out the components of the sanitation value chain along with steps taken by the state government under each head for bolstering FSSM on the ground. Furthermore, Chapter 4 taps into the institutional arrangements at the state, district and local levels for implementing sanitation-related initiatives along with capturing the efforts made for strengthening institutional arrangements and building capacities on FSSM. Chapter 4 concludes by addressing the progress made on the ground.

The lack of solid waste management (SWM) engenders diverse environmental and health risks across spatial and temporal scales. It has been a longstanding issue nationally and three out of ten households in urban Odisha too reported dumping their solid waste in the environment in 2018. Yet, within a couple of years, the state has demonstrated remarkable and fast-scaling success in SWM, with door-to-door collection of source-segregated solid waste soaring to nearly 100% across all cities and towns in the state. Chapter 5 on *Municipal Solid Waste Management with Community Partnership* charts the state's trajectory in developing and scaling a unique model for urban SWM that valorises decentralisation, local ingenuity, circular economy and community partnerships.

Chapter 6 on *Tap Water Supply at Every Home and Beyond* presents an account of the various initiatives undertaken by the HUDD to achieve the goal of ensuring universal access to reliable and good quality water supply for all urban households in the state. It describes the situation prior to the implementation of the initiative, with low coverage, uneven availability and poor quality of piped water supply being the norm. Chapter 6 details the approach adopted for implementation of the programme, including (a) setting well-defined time-bound service delivery targets and (b) phase-wise implementation, ensuring that around 800 water supply projects were executed within a short period of time. Furthermore, Chapter 6 documents the key innovations of the programme that have been crucial for success and the impact of the initiative on water service delivery performance indicators – including reduction of infections and deaths due to water-borne diseases. Chapter 6 concludes by presenting an overview of the future vision of the state government in the urban water sector.

Chapter 7 on *AAHAAR: An Affordable Meal Programme* documents Aahaar – an affordable meal programme for the underprivileged being implemented by the HUDD. At the outset, Chapter 7 describes the initiative's aims of honouring the obligations of the state with respect to the right to food as well as the UN Sustainable Development Goals relating to 'zero hunger'. It goes on to discuss the steps adopted for designing the programme as well as its approach and the value chain. Subsequently, it outlines the five cornerstones of the Aahaar programme which make it a truly inclusive, accessible one providing affordable, nutritious and wholesome meals. Chapter 7 traces the phase-wise implementation of the programme, which began as a pilot initiative covering five Municipal Corporations and subsequently expanded to cover 158 towns along with 56-night Aahaar centres. Furthermore, it presents the innovative aspects of the programme including ISO certification, leveraging funds from multiple sources, and creation of local institutions at city and district levels for programme implementation and monitoring. Chapter 7 closes with an account of the impact of the initiative.

To ensure inclusive development and recognising the significant contribution of the slum dwellers to the city's economy and growth, the Odisha government enacted the Odisha Land Rights to Slum Dwellers Act, 2017, and expanded it with the Odisha Liveable Habitat Mission or JAGA Mission in 2018. It aimed to assign land rights to eligible slum dwellers for redevelopment, rehabilitation and upgradation of slums across the state. Backed by innovative approaches such as the use of drone technology, community mobilisation and exploring land tenure continuum, among others, Odisha could provide tenure security to around 60 000 slum households. The JAGA Mission has achieved global recognition, winning prizes at the World Habitat Awards 2019 and 2023. It had also won the India Geospatial Excellence Award for technological innovation in empowering urban poor by providing land rights. Chapter 8 on *Land Rights for Slum Dwellers: Jaga Mission Phase 1* details the trigger, enabling environment and the detailed procedure for distributing land rights to the slum dwellers.

In its second phase, the JAGA Mission moved towards transforming the slums into liveable habitats called Adarsh Colonies. While the government committed to securing land rights for the urban poor in the first phase of the JAGA Mission, in the second phase, it prioritised slum upgradation and delisting through creation of necessary basic civic infrastructure and bringing the tenable slums at par with the rest of the city. However, there remained people at the margins of the cities occupying environmentally hazardous land or living under significant threat of eviction. In this second phase, towards securing residents' 'right to the city' and pursuing the spirit of 'leave no one behind', the Government of Odisha has also undertaken relocation to new habitats, by sites and services programme. The second phase has continued to explore strategies for converting de facto to de jure land title, to ensure tenure security to slum dwellers on untenable land. With the willingness of the government and the support of the community Odisha could bring around 550 slums at par with the cities and delist them. Chapter 9 on *Right to Basic Amenities and Services to the Slum Dwellers: Jaga Mission Phase 2* focuses on procedural details as well the enabling environment for effective implementation of slum upgradation and delisting as well as creating a new habitat.

Chapter 10 on the *Seeding Social Change by Underwriting Dignity and Safety of Sanitation Workers: Garima Scheme* details the uniqueness of the scheme introduced by the Government of Odisha for safety and dignity of core sanitation workers in 2020. It is the first scheme of its kind launched by any state in India. It will be implemented in all ULBs of Odisha. The scheme provides an enabling environment to regulate core sanitation services. At the same time, it offers a comprehensive package that ensures service level benefits, social security and financial benefits to core sanitation workers and their families. This scheme would cover nearly 20 000 core sanitation workers and their families. It is likely to benefit almost 0.1 million population.

The COVID-19 pandemic has had a significant impact on economic activities across the country. In response to the rising unemployment and vulnerability of informal labour and migrants, the Odisha government launched the Urban Wage Employment Initiative (UWEI) on 18 April 2020 across the 115 ULBs of Odisha. Odisha was the first state to launch an urban wage employment initiative during the pandemic. Chapter 11 on *Employment Safety Net for The Urban Poor: UWEI/MUKTA Scheme* focuses on the salient features and implementation of UWEI in Odisha. Chapter 11 discusses the state's innovative governance model in which community-based organisations have emerged as the fourth tier of governance that has improved the creation as well as maintenance of community infrastructure in the urban areas.

Chapter 12 on *Other Flagship Initiatives of HUDD, Government of Odisha* showcases additional flagship initiatives introduced by the Government of Odisha to further advance urban transformation in the state. It also delves into the role of the Odisha Urban Academy in propelling the change agenda forward. The Academy achieves this by spreading new, practice-based knowledge and crafting innovative training solutions for a variety of stakeholders. These efforts are geared towards enabling and supporting transformative changes throughout the urban landscape.

Odisha has demonstrated that partnership between citizens, municipalities and the state can deliver transformative urban change in the state. The Slum Dwellers' Associations are emerging as the fourth tier of governance structures with urban poor communities actively participating in the development and governance process. The Odisha model of decentralisation is unique, as it is conceptualised, designed and executed as an integral part of programmes/ schemes. It makes the partnership process compulsory and not optional. Mission Shakti Groups, Slum Dwellers' Associations, Transgender Groups, Ragpickers' Group, and so on are now active partners of ULBs and function as the fourth tier of the governance system. Chapter 13 on *Community Partnerships Towards Building a Fourth Tier of Urban Governance* aims to bring out the key enabling decisions by the government that have resulted in capacity building, institutional strengthening and economic empowerment of the community.

Chapter 14 on *Key Facets of the Odisha Urban Development Model* discusses the state's overall strategy for ensuring inclusive growth and socio-economic empowerment of all citizens, through partnerships, collaborations and joint initiatives of the government, civil society organisations, academics and corporates. It further summarises the learnings from the state's experience of implementing various urban development programmes as discussed in different chapters of the book and points out how other Indian states can adopt these learnings for their respective urban development programmes.

REFERENCES

Bercegol R. (2017). *Small Towns and Decentralisation in India*. Springer.

Kundu A. (2007). Migration and urbanisation in India in the context of poverty alleviation. In International conference and workshop on Policy Perspectives on Growth, Economic Structures and Poverty Reduction. Beijing, China. Retrieved from https://smartnet.niua.org/sites/default/files/resources/Urbanization%20and%20Migration_Kundu.pdf

Mathur O. P. (2013). Toward a System of Municipal Finance for Twenty-first Century India. *New Century Local Government: Commonwealth Perspectives*, 91.

Sharma C. K. (2014). Governance, governmentality and governability: Constraints and possibilities of decentralization in South Asia.

Shastri S. and Singh K. (2023). Emerging trends in Indian federalism: Collaboration, competition and coordination. *Cooperative Federalism in South Asia and Europe*, 104–119.

doi: 10.2166/9781789064926_0009

Chapter 2

Urbanisation and economic growth

ABSTRACT

This chapter unpacks the trajectory of Odisha's urbanisation, along with the demographic, economic and social aspects that have shaped the same. Some of the key points discussed in this chapter include the state's population growth trends, urban growth pattern, namely, spatial distribution of towns, urban inequalities and slum formation, economic indicators such as gross state domestic product, per capita income, workforce participation rate.

Keywords: Urbanisation, migration, economic growth, spatial inequality

2.1 URBANISATION: NATIONAL TRENDS AND PATTERNS

Urbanisation and economic growth have gone hand in hand since the beginning of civilisation. How nations have handled population growth in their urban centres has become a determining factor in keeping their economic growth on the ascendant. This is also true for India, which after millennia of remaining predominantly rural, has started seeing an era of sustained urban and economic growth.

The growth of industrial cities in the late 19th and early 20th centuries marked the introduction of modern urban governance structures in the country. In the 1990s, the pace of urban expansion rapidly accelerated in India. Urban population rose from 285 million in 2001 to 377 million in 2011. This has been the critical driver of the economy, considering that approximately one-third of the population accounts for 63% of India's gross domestic product (GDP). Both the urban population and its share in the economy are expected to grow, with projections that urban India will generate 75% of the GDP by 2030 (CBRE, 2019).

Economics aside, urban areas are also expected to host more than 50% of the country's population by 2050 (United Nations, 2022). This rapid rate of urbanisation will exert significant pressure on all sorts of urban infrastructure, ranging from housing and water to sanitation and resilience. All this in the face of a low land-to-population ratio that has declined compared to 1960 (Kapur *et al.*, 2014).

The high rate of urban growth is also not evenly distributed across India, with urbanisation more rapid in some regions than others. Each state faces a different set of challenges and contexts in ensuring availability and participation of workforce for viable clustering of secondary and tertiary production functions.

2.2 ODISHA'S URBAN LANDSCAPE

Located on the east coast of the subcontinent, Odisha is one of the least urbanised states in India. In 2011, the state's population was 41.9 million (Census of India 2011), which accounted for about 3.5% of the national population. Of this, only 17% live in urban areas, well below the national average of 31% as shown in Figure 2.1.

Odisha's urban system presently (in 2021) contains 115 urban local bodies (ULBs), segmented as five municipal corporations (MCs), 48 municipalities and 61 notified area councils (NACs).

Between 2001 and 2011, the state's urban population rose by two percentage points annually, with a corresponding drop in the rural share. This process has

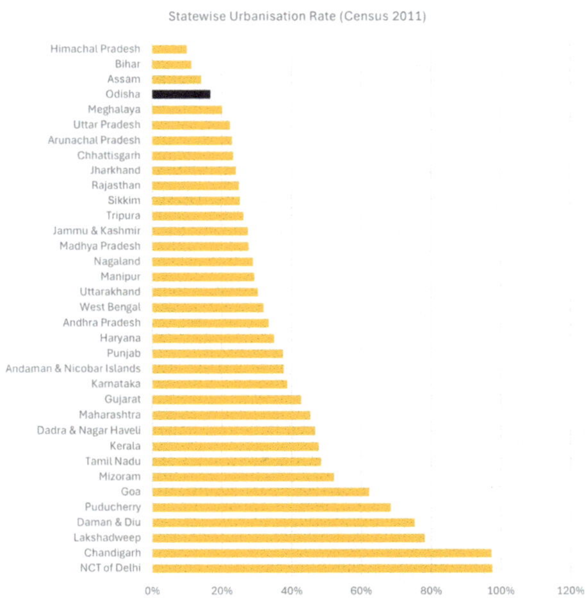

Figure 2.1 Odisha's urbanisation. *Source*: Census of India 2011.

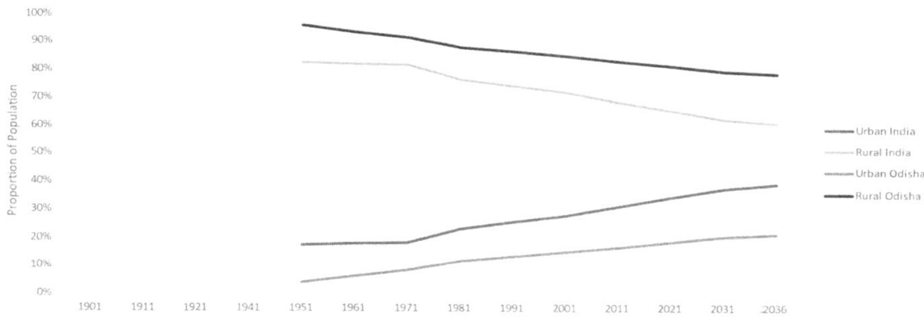

Figure 2.2 Rural and urban population growth over time in Odisha and India. *Source*: Census 2011, World Urban Projections (2018–2021) and Report of the Technical Committee on Population Projections, 2019 (Population projections for India and states, 2011–2036).

only accelerated since. By 2036, urban Odisha is expected to comprise 22% of the population, a significant change in character from the 4% of the 1950s as shown in Figure 2.2.

Given the distribution of the population across the state's geography, academic discourse on Odisha has revolved around the concepts of inadequacy, poverty and backwardness, underdeveloped agriculture and social deprivation (Mitra, 2019). But the sheer scale of urbanisation in the state in more recent times may well lead to a turnaround in the discourse, presenting a rare opportunity for advance management of both the process (of urbanisation) and its repercussions, through the mobilisation of resources and capacity.

The (in situ) nature of urban expansion in Odisha also presents an opportunity to move towards sustainability and inclusivity, in line with the United Nations (UN) sustainable development goal (SDG) 11, while also building systemic resilience against climate change in a state that frequently bears the brunt of highly destructive cyclones.

There is also a geographical skew in the loading of urbanisation rates across the state. More than half of the urban population (52%) is concentrated in districts that cover just 20% of the total land area. These districts with the highest levels of urbanisation are mainly in the eastern belt, along the 480 km coastline of the Bay of Bengal, attributable to these regions having better road connectivity, industrial development and rising employment prospects. The disparity in urbanisation is also evidenced by the high levels in Khordha district (48%) against less than 5% in Boudh, in central Odisha as shown in Table 2.1.

---

[1] Constitution (Seventy-Fourth Amendment) Act, 1992 has introduced a new Part IXA in the Constitution, which deals with Municipalities in an article 243 P to 243 ZG. This amendment, also known as Nagarpalika Act, came into force on 1st June 1993. It has given constitutional status to the municipalities and brought them under the justifiable part of the constitution. States were put under constitutional obligation to adopt municipalities as per system enshrined in the constitution.

Table 2.1 Urban population across city class size.

District	Proportion of Urban Population 2011 (%)	No. of Class 1 Cities	No. of Class II–VI Cities
Khordha	48.1	1	4
Jharsuguda	39.9	0	3
Sundergarh	35.5	2	4
Sambalpur	29.8	1	4
Cuttack	27.9	1	3
Ganjam	21.8	1	17
Koraput	16.4	0	4
Anugul	16.2	0	3
Puri	15.6	1	3
Rayagada	15.4	0	3
Keonjhar	14.0	0	4
Bhadrak	12.3	1	2
Gajapati	12.2	0	2
Balangir	12.0	0	4
Balasore	10.9	1	4
Jagatisinghpur	10.2	0	2
Bargarh	10.1	0	3
Kandhamal	9.9	0	2
Dhenkanal	9.8	0	3
Nayagarh	8.3	0	2
Malkangiri	8.0	0	2
Kalahandi	7.8	0	3
Mayurbhanj	7.7	1	3
Sonepur	7.7	0	3
Jajpur	7.4	0	2
Nabarangpur	7.2	0	2
Deogarh	7.2	0	1
Kendrapara	5.8	0	2
Nuapada	5.6	0	2
Boudh	4.6	0	2

Source: Census of India 2011.

2.3 URBANISATION, BOTTOMS UP

Odisha has also witnessed high levels of subaltern urbanisation, with the bulk of growth occurring not in the larger cities but in the smaller ones. Towns have expanded rapidly and small urban pockets have mushroomed across the state. These pockets display urban characteristics though they may not be formally classified as such. It is these Class IV, V and VI cities that have been the engine of urban expansion in Odisha over the last five decades as shown in Figure 2.3.

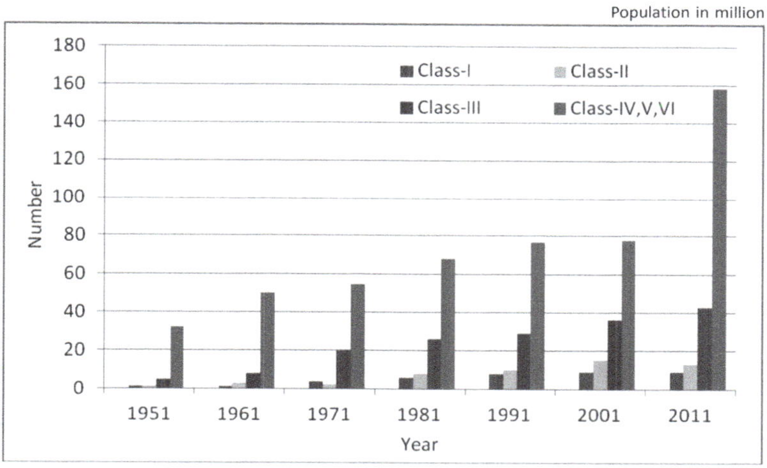

Figure 2.3 Population size classes. *Source*: State Action Plan on Climate Change 2018–2023.

This small town-led growth pattern of Odisha has led to certain imbalances in financial resources, capabilities in spatial planning and administrative bandwidth, to name a few. Numerous scholars have contended that the incomplete implementation of the 74th Amendment to the Constitution[1] has given rise to a common set of problems across the board nationally.

Small towns often face difficulty in finding sources of revenue or are limited by weak institutional capacities in collecting these revenues. As a result, the already limited finances available are exhausted on salaries and maintenance, leaving little to no bandwidth for developmental work. The boom in small towns has also led to increased pressure on land, housing and the provision of civic services, reinforcing the need for more strategic approaches to urban planning and development.

2.4 UNDERSTANDING INTERNAL MIGRATION

Odisha's rising urbanisation is driven by internal migration from rural areas into urban pockets in search of better economic opportunities and higher quality of life, according to the Odisha Economic Survey 2020–2021 (Government of Odisha, 2021). The decadal urban growth rate in the 2001–2011 period (26.9%) was far higher than the overall population growth of 14% as shown in Figure 2.4. Of the 7 million people who live in urban areas, over 2 million are estimated to have moved here in the 2001–2011 period.

According to data from the National Sample Survey (NSS) 64th round of 2008, 4.3% of urban households in Odisha was classified as migrant households, which means they had moved to the urban locality sometime in the previous 365 days as shown in Figure 2.5. The phenomenon of high levels of migration from rural to urban areas prevails across the state, pointing to a movement away from agriculture and towards industries and services as shown in Figures 2.6 and 2.7.

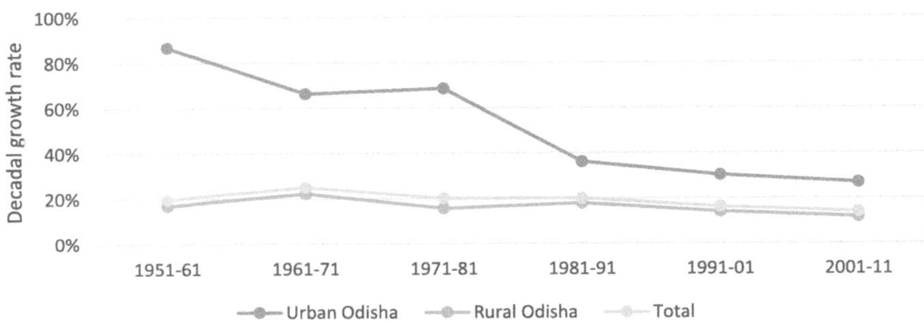

Figure 2.4 Trend of decadal growth rate of Odisha. *Source*: Census of India 2011.

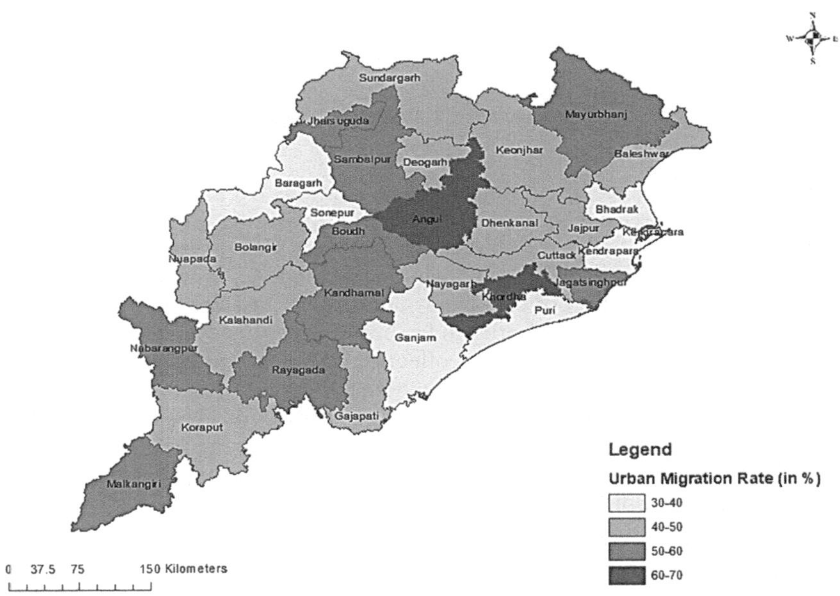

Figure 2.5 Rate of migration. *Source*: National Sample Survey (NSS) 64th round of 2008.

There was a crucial fallout from this high rate of migration during the outbreak of the COVID-19 pandemic in 2020 and the subsequent lockdown, which saw most economic activities come to a grinding halt. Particularly badly hit by job losses across the country were daily wage workers and casual laborers. With Odisha being one of India's biggest sources (origin) of internal migration, the lockdown triggered a huge wave of workers returning en masse.

Between April and September 2020, more than 1 million migrants returned to Odisha from other parts of the country, with the bulk of them coming from

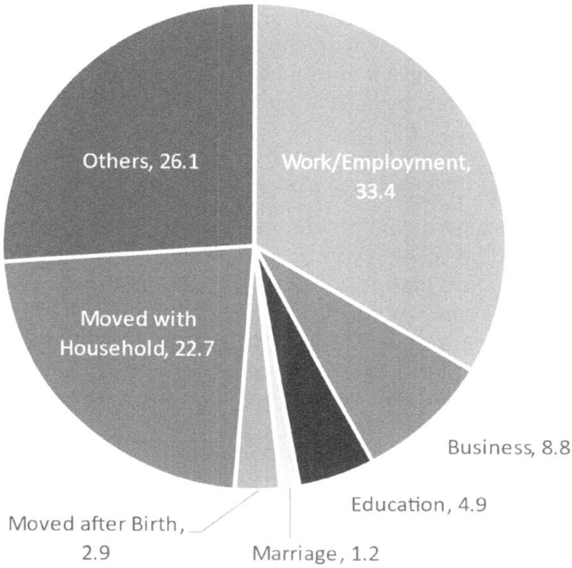

Figure 2.6 Reasons for migration to urban areas among males. *Source*: Census of India 2011.

Figure 2.7 Reasons for migration to urban areas among females. *Source*: Census of India 2011.

Gujarat, Tamil Nadu and Kerala. The lockdown also triggered a large-scale movement of people within the state, with migrants returning from Odisha's cities to their villages and hometowns. A significant number among those who returned were left with few options of livelihood during and soon after the lockdown.

2.5 ODISHA'S DEMOGRAPHIC DIVIDEND

Odisha's urban development needs are closely related to its demographic strengths. The state's young and rapidly growing urban population is the demographic dividend that is vital to its economic growth and prosperity for decades to come. According to Census 2011, 15.6% of Odisha's population is under 9 years of age, and a whopping 44.6% under 25. Crucially, 67% of the population is in the working age group (15–59 years). Furthermore, the median age of Odisha is 26.6, similar to the national median age of 24.9 as shown in Figure 2.8.

The demographic dividend potential of urban Odisha, if harnessed, could prove beneficial to its cities and towns. Expanding value addition chains and tertiary economies hold promise for a young workforce that stands to benefit economically from the creation of opportunities.

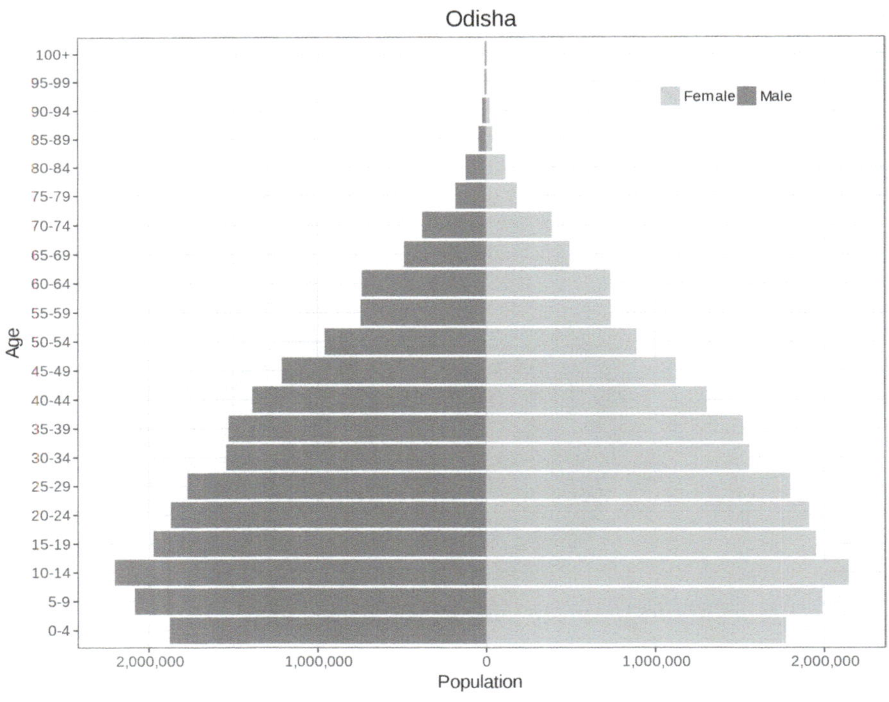

Figure 2.8 Population pyramid of urban Odisha. *Source*: Census of India 2011.

Odisha will be able to derive the benefit of its demographic dividend only if it can ensure that the employability of its workforce remains high. For this, it will need to create systems that focus on capacity creation, skilling, health, education and vocational training. Apart from this, it will also have to put in place appropriate land and housing policies in conjunction with good governance.

2.6 URBAN SPATIAL INEQUALITY

The rising level of inbound migration from rural areas has led to a shortage of affordable space in the cities. The result is the growing spread of slums and squatter settlements, which are marked by inadequate housing and lack of basic sanitation and hygiene facilities, raising concerns over health, education and standard of living. These deficiencies could have a detrimental effect on the mobility and skill levels of the workforce, affecting Odisha's ability to take advantage of its demographic dividend.

As per Census 2011, about 1.56 million urban dwellers (800 963 males and 759 340 females) resided in slum areas, making up about 23% of the state's urban population. Slum populations are critical to the health of the urban economy, and make vital – and often undervalued – contributions to economic activity and growth. This is reflected in the work participation rate (WPR), which is the proportion of workers to total population. The WPR of Odisha's slum population is 36%, which is similar to its urban WPR.

The incidence of slums is evidently widespread in Odisha. Large cities like Bhubaneswar and Cuttack recorded the highest number of slums with populations of 0.16 million each. The 10 Class I cities (population over 100 000) contained 52.6% of the slum population of the state. The rest lived in 65 of 97 Class II–Class VI cities as shown in Figure 2.9.

While crucial to the urban economy, the rising slum population in the state also serves to underline trends in urban poverty. This is a vulnerable section of the population that has to contend with insecure and inadequate housing and civic services, and limited access to services like health, nutrition, water and sanitation.

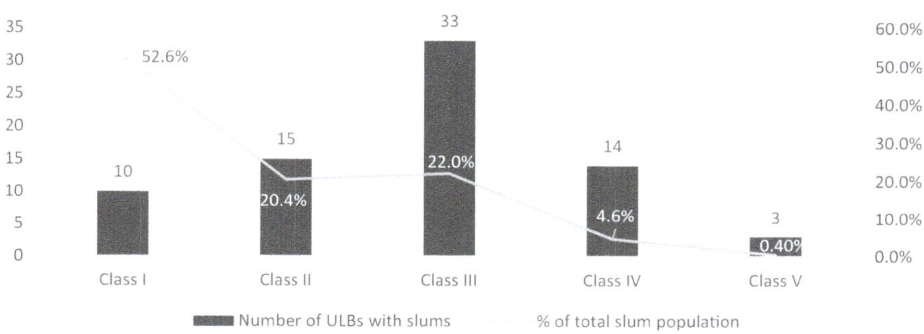

Figure 2.9 City class-wise distribution of slum population. *Source*: Census of India 2011.

According to the 2011 Census, 48% of the state's 0.35 million slum households had to rely on open defaecation. More than half did not have drainage connectivity and 39% accessed water through hand pumps, tube wells and bore wells. These problems, combined with overcrowding, lead to a heightened risk of ill-health and resultant loss of opportunity.

The small town-led growth of urbanisation in Odisha has led to the formation of slums in these areas as well. While this is of concern, there is also a unique opportunity to address common urban issues like land tenure, housing and service delivery even as these slums are formed.

2.7 FUELLING ECONOMIC GROWTH

A strong urban economy is the driver of development. India's urban population of 31% accounts for about 60% of its GDP. A collaborative research study by the Human Settlement Management Institute (HSMI) and National Institute of Urban Affairs (NIUA) in 2014 established high positive inter-linkage effects between urbanisation and per capita gross state domestic product (GSDP) as shown in Figure 2.10.

India's Twelfth Five-Year Plan (2012–2017) too noted that urbanisation would be central to India's strategy of achieving faster and more inclusive growth (Planning Commission, GoI, 2013). It reasoned that agglomeration and densification of economic activities stimulate economic efficiencies and provide more opportunities for earning livelihoods, thereby enabling faster inclusion of increasing numbers of people in the process of economic growth. Harnessing this would remain critical to achieving and sustaining a high-growth path.

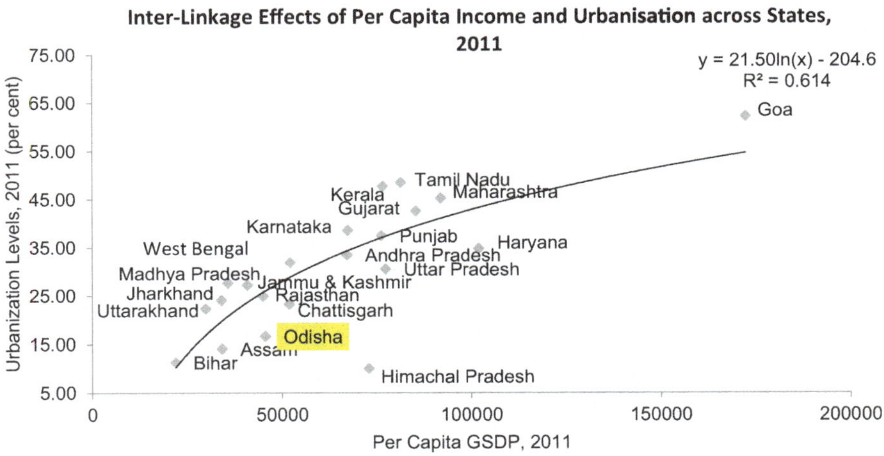

Figure 2.10 Interlinkage effects of per capita income and urbanisation across states, 2011. *Source*: https://smartnet.niua.org/sites/default/files/resources/HUDCO-Consolidated.pdf

		Employment Share (in per cent)			
		1993-94	2004-05	2011-12	2019-20
Agricultural	Rural	72	69	62	55
	Urban	10	14	14	5
Industry	Rural	9	17	23	26
	Urban	26	27	23	33
Services	Rural	18	14	15	19
	Urban	65	60	63	62

Figure 2.11 Employment share in Odisha. *Source*: Economic Survey 2020–2021.

Odisha has reflected similar trends, with its rising urbanisation contributing to the transformation of its economic outlook.[2] In just 2013, the Rajan Committee had termed Odisha 'one of the least developed States in the country' (Government of Odisha, 2021). The state also recorded the fifth highest growth rate in the country between 2011–2012 and 2019–2020, ahead of 13 others. If such trend continues, the state will be well on its way to move away from the tag of least developed States in the country.

According to the state's Economic Survey 2020–2021, a comparative analysis of the growth performance of Odisha and other states during 2012–2013 to 2019–2020 reveals a positive differential trend in the state's average annual growth rate. It indicates that in this period, Odisha's economy grew at an average annual rate of 7.1%, which was faster than the national average of 6.6% as shown in Figure 2.11.

Odisha's GSDP at current basic prices increased from Rs 2 309 878 million in 2011–2012 to Rs 3 928 037 million in 2016–2017: an increment of 70% over a span of five years. The high GDP growth during this period may be attributed to higher agricultural productivity, fiscal consolidation and an increase in manufacturing output. GSDP rose to Rs 5 212 746 million in 2019–2020, a 33% surge over 2016–2017 and 126% over 2011–2012. GSDP at constant prices (2011–2012) jumped from Rs 2 309 878 million in 2011–2012 to Rs 3 964 994 million in 2019–20 (Government of Odisha, 2021) as shown in Figure 2.12.

While the pandemic dealt a huge blow to India's GDP, there are indications of a declining trend over the last decade, especially after 2016–2017. This was exacerbated by demonetisation and the goods and services tax (GST), which shrank GDP growth from over 8% in 2016–2017 to 4% in 2019–2020, even before COVID-19 set in. Odisha did not escape the adverse impacts of the COVID-19 pandemic. According to advance estimates of GSDP at constant prices (2011–2012), it fell to Rs 3 769 984 million in 2020–2021. This is a contraction of 4.2% against the national estimate of 7.7%.

[2] https://odisha.gov.in/sites/default/files/2022-03/Economic_Survey_2021-22_0.pdf

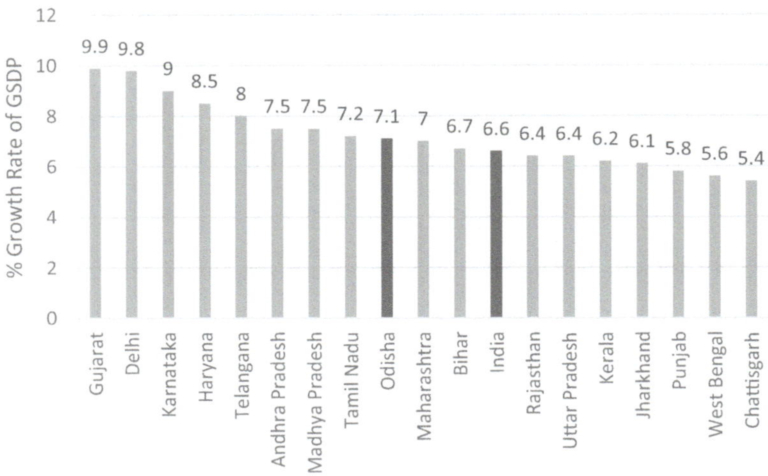

Figure 2.12 Average growth rate of GSDP of major states vs India. *Source*: Economic Survey of Odisha 2020–2021.

2.8 MAKING STRIDES IN URBANISATION, WINNING AWARDS

Recognising the state's unique opportunity for urban development, the Government of Odisha has already achieved significant milestones in terms of both structure and implementation. Strides have been made in slum development, provision of adequate water and sanitation, education and public health infrastructure. For example, 727 slums have been upgraded since 2019 with eight cities having been declared slum free, household tap connections increased from 0.3 million in 2017 to 1 million in 2021, and treatment of faecal waste has jumped from 2% in 2015 to 98% in 2022 as per officials of HUDD, Government of Odisha. The state government is now building on these successes to amplify efforts for further improvements in urban infrastructure, quality service delivery, decentralised governance mechanisms and community-driven programmes.

The state's welfare-driven approach has been recognised through numerous awards and accolades for its urban development initiatives. Odisha was adjudged the 'Best State' and the 'Best Civic Agency' awards at the second Janaagraha City Governance Awards 2021, which commemorate ground-breaking work in decentralisation in the urban context. Odisha's ambitious Jaga Mission, the world's largest slum land titling and habitat programme, is also the only initiative to have won the World Habitat Award twice – first in 2019 and then in 2022. Jaga Mission was also awarded the Asia-Pacific Housing Forum Award for Innovation in Public Policy in 2021.

The state's efforts in providing piped water supply are also widely appreciated. It won the Award of Excellence at the 18th CSI SIG e-Governance Award 2020 for its Drink from Tap Mission and Smart Water Management in the temple town of Puri. Similarly, WATCO was selected for the Global Water Leader Award 2022 held in Spain.

Furthermore, the Urban Wage Employment Initiative (UWEI) implemented across the state and the Faecal Sludge and Septage Management (FSSM) programme in Berhampur have been shortlisted among the top 15 urban development projects at the prestigious 2020 Guangzhou International Awards for Urban Innovation, as exemplary models of innovative policies and practices. UWEI has also been selected as the top five finalists for the WRI Ross Center Prize for Cities 2021–2022.

Over the last few years, the Government of Odisha has introduced multiple urban development initiatives, as the book's subsequent chapters will outline. These include programmes and initiatives in piped water supply, faecal sludge management, solid waste management, slum upgradation and delisting, and urban wage employment.

Emerging as a pioneer in implementing successful, welfare-oriented urban interventions, the state has made notable progress in managing urbanisation. The pro-poor urban development agenda of the government has transformed Odisha into a leading state in encouraging sustainability, equality and inclusivity in its cities.

REFERENCES

CBRE (2019). India 2030 – Exploring the Future. Retrieved from https://credai.org/console/public/upload/4de5a183620d1ae193c813f2d20c7dec.pdf

Census of India (2011). Office of the Registrar General & Census Commissioner, Government of India.

Government of Odisha (2021). Odisha Economic Survey 2020–21. Retrieved from http://www.desorissa.nic.in/pdf/Odisha%20Economic%20Survey%202020-21-1.pdf

Kapur D., Somanathan T. and Subramanian A. (2014). Land-shackled–I. *Business Standard*. Retrieved from https://www.business-standard.com/article/opinion/devesh-kapur-t-v-somanathan-arvind-subramanian-land-shackled-i-114072000708_1.html

Mitra P. P. (2019). Odisha: Analysing Some Facets of the State Development Strategy. Indian Journal of Public Administration. Retrieved from https://journals.sagepub.com/doi/pdf/10.1177/0019556118820405

Planning Commission, GoI. (2013). Twelfth Five Year Plan (2012–2017). Sage. SAGE Publications India Pvt Ltd, India. Retrieved from https://nhm.gov.in/images/pdf/publication/Planning_Commission/12th_Five_year_plan-Vol-1.pdf

United Nations (2022). *World Population Prospects 2022*. Retrieved from https://population.un.org/wpp/

doi: 10.2166/9781789064926_0023

Chapter 3

Odisha's devolved, innovative urban governance

ABSTRACT

This chapter outlines the vision, mission, functions and objectives of the Housing & Urban Development Department (HUDD) and its directorates (attached offices). The purpose of this chapter to map the governance framework for urban Odisha, establishes the functionalities of the agencies under the aegis of the HUDD, outline the hierarchical structure of the officers, to better describe the roles and duties of each actor at the state/district/ULB level in the implementation of the programmes discussed in the book.

Keywords: Urban governance, urbanisation, governance framework, reform

3.1 BACKGROUND

Urbanisation brings with it a progressively rising demand for resources and services. Rising rates of urbanisation, especially in the developing world, have underscored the need to build up strong institutional frameworks and service delivery mechanisms. Urban governance is a challenge as it is for administrators and planners, given the ever-changing populations, economies, demographics and land-use patterns. But globalisation and its demands for competitiveness have compounded the under-preparedness of cities to remain both inclusive and economically successful.

Urban governance reforms are gaining in importance as cities in the Global South look to manage their myriad challenges. In India, the push to decentralise urban governance and devolve powers came in the form of the **74th Amendment** to the Constitution in 1992. This landmark legislation formally recognised ULBs as the third tier of government along with the central and state governments as first and second tiers. The 74th Amendment listed under **Schedule 12** several functions that were to be the responsibility of local bodies: such as urban planning, including town planning; regulation of land use; construction of buildings, roads and bridges; and the provision of water, public health, sanitation and solid waste management.

However, while the legislation mandated devolution of powers, it remained largely silent on questions of finance and resource mobilisation. State governments were tasked with allocating functions with funds and resources, and to provide an enabling environment through legislative and institutional means. The central government was tasked with strategic leadership. These systemic characteristics led to a perilous state of municipal finances, which have been worsened by the COVID-19 pandemic.

Given the critical situation, there is an urgent need to integrate reforms in urban governance with innovations in the field to address the growing complexities in the urban sector. Successful service delivery depends on how well state and local governments can work collaboratively; understanding how innovative schemes work also becomes critical for them to be replicated in other urban areas. Odisha's ongoing evolution in urban governance, particularly in institutional frameworks, is worth examining as a model for elevating community participation, inclusiveness and decentralisation in urban governance.

3.2 DRIVING FROM ABOVE: ODISHA'S DEPARTMENT OF HOUSING & URBAN DEVELOPMENT

Starting from a base of low urbanisation, Odisha has over the past decade achieved significant strides in urban governance. This is in large part thanks to a spectrum of developmental initiatives launched and shepherded by the Housing & Urban Development Department (HUDD), the state's nodal unit for planned urban growth. These not only raise the profile of governance in the existing cities but also prepare them for the cities they are set to become.

HUDD is responsible for urban governance, planning and delivery of services like affordable housing, safe drinking water, sanitation, solid waste management, storm water drainage, sewerage, roads, public transport, positioning of cities for economic growth, and increasing employability of the urban poor. The state's minister for Housing & Urban Development heads HUDD, and it is administratively led by a Principal Secretary rank officer of the Indian Administrative Service.

Over the past decade, HUDD's efforts have focused on the formulation of sustainable policies and ensuring their successful implementation. In tandem there has been a drive to improve efficiency by monitoring, transparency, fostering inclusivity through community participation, raising capabilities of local bodies and parastatal agencies, and a people-oriented approach to the delivery of civic services.

This successful implementation of a broad spectrum of schemes has led to dramatic improvements in Odisha's performance across numerous developmental indices. The state's performance in the achievement of SDGs was 61 in 2020, up 10 points from the previous year. On SDG 11, which is to do with sustainability of cities and communities, Odisha has emerged as a front runner on the SDG-India Index & Dashboard 2021 with a composite score of 70 points (NITI Aayog, 2021).

Additionally, the state ranked first in 2020 on the Urban Governance Index (UGI) developed by the Praja Foundation (Praja Foundation, 2020). UGI ranks states on several themes focusing primarily on the structural aspects of urban governance, such as empowered local representatives, legislative structure, city administration, citizens and fiscal empowerment.

3.2.1 Vision and mission: urban vibrance

The cornerstone of HUDD's vision for Odisha's urban governance is the understanding that urban centres are the engines of economic growth. Hence, the lookout is bifocal: not limited to ensuring sustainable and well-provided urban spaces today but also gearing these cities to deal with increasing levels of economic success tomorrow.

To this end, the push has focused on inclusive cities with sustainable, efficient and flexible governance frameworks; economic opportunities for all sections; community participation; skilling of urban poor for enhanced employability; and the creation and maintenance of durable and relevant public infrastructure and civic services, including affordable housing.

HUDD's pivotal role in this mission is in ensuring the convergence of the efforts of a decentralised and empowered set of institutions and agencies, ranging from ULBs to community groups and parastatal agencies to stand-alone citizens.

3.2.2 Objectives

As per the official website[1] of the HUDD, the objectives are as below:

- Facilitating reforms and policy interventions to improve the financial base of ULBs and their interface with citizens in service delivery.
- Improving urban planning, governance and capacity development of ULB personnel for effective implementation of plans and rules, efficient urban management and service delivery.
- Enhancing service standards, choice of citizens, transparency, accountability and simplification of processes for grievance redressal.
- Encouraging public–private partnership (PPP) projects for urban infrastructure development and urban services, including urban transport.
- Effectively implementing affordable housing schemes for economically weaker section, lower income group, lower middle-income group and middle-income group.
- Effectively implementing programmes for urban poverty reduction.
- Effectively implementing and operating water supply, sewerage, solid waste management, storm water drainage and other urban development projects.
- Encouraging innovations in urban development.

---

[1] http://urbanodisha.gov.in/About.aspx

Table 3.1 HUDD's roles and responsibilities.

Urban Infrastructure Planning and Development	• Implementation of urban governance reforms (e.g. AMRUT, 14th Finance Commission, etc). • Formulation and implementation of enabling policies/guidelines. • Formulation and implementation of the Affordable Housing scheme with mandatory provision for creation of housing stock. • Implementation of Comprehensive Development Plans (CrDPs), City Development Plans (CDPs) and Master Plans (MPs). • Formulation and implementation of PPP policy on urban development. • Facilitating PPP projects in urban infrastructure, transport and solid waste management. • Designing and implementation of innovative pilot projects in selective areas with community participation.
Service Delivery Planning and Improvement	• Effective implementation of e-governance programmes. • Benchmarking urban services (Service Level Benchmarking – SLB). • Formulation and implementation of citizen's charter; public disclosure law. • Mobilisation of funds from various sources for successful implementation of flagship projects. • Implementation of replicable models/good practices.
Urban Poverty Alleviation	• Encouraging group productive activities of urban poor women for economic entitlement. • According to limited property rights to slum dwellers and prevention of new slums. • Dovetailing non-government/private/corporate partners for skill training and placement of youth.
Organisational Development and Capacity Building	• Capacity development of existing personnel and elected representatives. • Creation of a dedicated urban cadre. • Ensuring effective monitoring, reporting and review of projects, conforming to the project timelines.

Source: http://urbanodisha.gov.in/About.aspx

3.2.3 Roles and responsibilities

The roles and responsibilities of the HUDD, Government of Odisha is indicated in Table 3.1.

3.3 STRUCTURE: GEARING TO WORK

Odisha's strides, as evidenced by various indices, are primarily the outcome of its decentralised approach towards urban planning, land use and urban poverty alleviation, including its promotion of community involvement. The state government has been relatively successful in devolving both functionaries (such as Slum Dwellers Associations (SDA)s, Jal Saathis, Swachh Saathis, Swachh Karmis, women's Self-Help Groups) and funds to civic bodies. This has resulted in the creation of an institutional environment conducive to the execution of appropriate policies for public welfare.

The state's decentralised urban governance mechanism is streamlined into five directorates as indicated in Figure 3.1:

(a) Directorate of Municipal Administration (DMA)
The DMA is the apex administrative organisation of the Government of Odisha that regulates the functioning of all Municipal Corporations, Municipalities and Notified Area Councils in performing their regulatory and enforcement activities. DMA ensures that they adhere to the policies, procedures and guidelines provided by the government to achieve effective municipal administration in the state. It functions as an integrated directorate in the HUDD, and is headed by the Director, Municipal Administration.
The DMA is also involved in the monitoring and supervision of the development-oriented functions of ULBs. It interacts with several other departments and directorates – such as the Public Health Engineering Organization, Directorate of Town Planning, Orissa Water Supply and Sewerage Board (OWSSB) and so on – to enable seamless delivery of urban civic services to the citizens.
The DMA coordinates and supervises the activities of ULBs regarding municipal tax administration, financial management, infrastructure development, town planning, urban health and sanitation, environmental management, programmes for the urban poor, project execution, and so on. Based on these parameters, it conducts a performance evaluation of all ULBs on an annual basis.

(b) Directorate of Town Planning
The Directorate of Town Planning is the nodal agency for all urban planning-related activities in the state. Established in 1954 as the Town

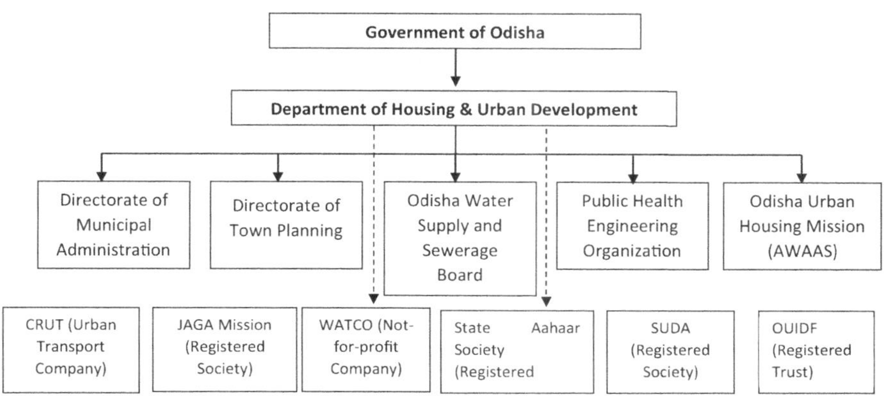

Figure 3.1 Institutional structure of HUDD. *Source*: author's analysis based on stakeholders' interview.

Planning Organisation, it was restructured as the Directorate of Town Planning in 1975. The Directorate has 20 Unit Offices at different district headquarters.

The Directorate monitors the preparation of Master Plans and Comprehensive Development Plans for the various planning areas of urban Odisha. The monitoring of different regions is carried out by the Directorate's Unit Offices, Special Planning Authorities and Regional Improvement Trusts.

The Directorate provides technical advice to the state government to aid the framing of policies for urban areas and to the Panchayati Raj Department to regulate the construction of large buildings in rural areas. Recently, it framed the Odisha Development Authorities (Planning and Building Standards) Rules which are a set of rules to universalise planning and building norms across the entire state; another aim is to automate the building plan approval process.

(c) Odisha Water Supply and Sewerage Board (OWSSB)

OWSSB was established in 1991 to ensure rapid development and proper regulation of water supply and sewerage services in the state. It is responsible for the execution of major water supply and sewerage projects. Once a project is completed, OWSSB hands it over to the Public Health Engineering Organisation (PHEO) for operation and maintenance. OWSSB is the nodal agency in the state for the implementation of urban sanitation projects. HUDD, through OWSSB, is implementing underground sewerage projects for Bhubaneswar, Cuttack, Puri, Sambalpur and Rourkela.

(d) Public Health Engineering Organization (PHEO)

The PHEO operates and maintains the water supply and sewerage systems of the urban areas of Odisha. To carry out the technical and administrative management of these systems, it utilised funds from the central and state governments and other sources.

(e) AWAAS – Odisha Urban Housing Mission (OUHM)

The AWAAS programme, also known as OUHM was set up with the goal of 'housing for all'. It was launched in 2015. Functioning as an integrated directorate under HUDD, it is also the State Level Nodal Agency (SLNA) for the implementation of central government housing schemes.

OUHM aims to create surplus housing stock through different strategic development models and ensure shelter for every identified homeless person in the state, including temporary migrants. This is to be achieved through the provisioning of permanent residential EWS (economically weaker section) and LIG (lower income group) units, as well as rental housing.

As a key programme of the government, OUHM is operated by a High-Level Committee (HLC), with the State Level Sanctioning and Monitoring Committee (SLSMC).

(f) Other verticals
- JAGA Mission: An ambitious scheme to provide housing and land rights to slum dwellers, with a focus on slum upgradation and delisting (detailed in Chapters 7 and 8).
- Odisha Water Corporation (WATCO): A not-for-profit company of the Government of Odisha (detailed in Chapter 5).

3.4 CONCLUSION

While the 74th Amendment to the Constitution attributes specific functions to various government departments, in practice, there are different institutions responsible for different parts of the same function. Ultimately, it is the ability of diverse arms and levels of the government to work together that ensures good governance and is key to the achievement of the New Urban Agenda (NUA)[2] adopted at the United Nations Conference on Housing and Sustainable Urban Development (Habitat III) in 2016.

Not just parts of the government, local bodies also need to scale up their engagement and cooperation with the private sector, civil society groups, and more importantly, the community. Further, a comprehensive roadmap to good urban governance hinges on the principles of transparency, inclusivity, decentralisation and accountability. Odisha's HUDD has successfully demonstrated a robust decentralisation process in its urban governance framework spanning key services such as water supply, sanitation, housing, employment and health.

[2] The New Urban Agenda represents a shared vision for a better and more sustainable future. If well-planned and well-managed, urbanisation can be a powerful tool for sustainable development for both developing and developed countries.

doi: 10.2166/9781789064926_0031

Chapter 4

State-wide environmental sanitation through innovations in faecal sludge and septage management

ABSTRACT

This chapter essentially speaks about the policy interventions and projects undertaken by Government of Odisha in the faecal sludge and septage management (FSSM) sector for ensuring environmentally sustainable sanitation. It provides a brief analysis of the situation existed before the FSSM journey which started in 2016 and the transformation that has happened in desludging and cesspool services and treatment facilities in both urban and rural areas.

Keywords: Co-treatment, decentralised wastewater system (DEWATS), self-help groups, appropriate technology, desludging, cesspool

4.1 OPEN TO SKY: A SEVERE LACK

Similar to other states in India, Odisha also suffered from deficiencies of sanitation facilities until 2016. Poor access to sanitation was compounded by the lack of adequate infrastructure and systems to address sewage and faecal sludge. This often meant that untreated waste was finding its way into the environment and water bodies, including the state's major rivers.

The Census in 2011 showed that about 65% of Odisha's urban households had individual household latrines (IHHLs). With just 2% of households using community toilets or public toilets, a worrying 33.2% reported they were defaecating in the open. This made Odisha rate of open defaecation in urban areas the second highest among Indian states (Figure 4.1).

Of the 65% households that had IHHLs, only 11.5% were connected to a sewer system but without any sewage treatment facility. For more than half of

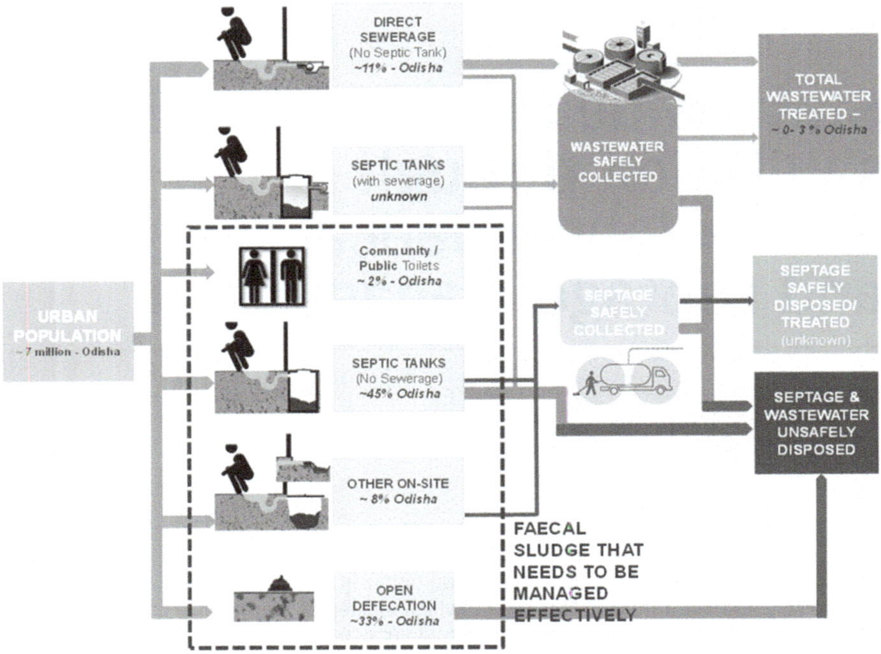

Figure 4.1 Faecal sludge flow diagram for Odisha. *Source*: https://www.indiawaterportal.org/articles/improving-city-wide-sanitation-service-delivery-small-towns-odisha.

the households, septic tanks and other types of on-site sanitation (OSS) systems such as pit latrines were the only options.

The prevalent practices in the state's urban areas indicated serious deficiencies in the construction of containment structures, irregular emptying of these facilities and inadequate desludging and cesspool services for collection and transportation of septage. Furthermore, only 2% of the faecal sludge being generated was being treated in sewage treatment plants (STPs) until June 2015.

The acute need to make faecal sludge and septage management (FSSM) the core of any intervention in sanitation was recognised at the highest levels of the state government. In other words, where waste would go from a flushed toilet was as critical as having a toilet itself.

4.2 SYNERGISING INTERVENTIONS: ODISHA'S FSSM JOURNEY

In Odisha's quest for improved sanitation through FSSM, the goal was to devise a system that would be relatively simple to construct, effective in treating faecal sludge while adhering to the quality and pollution control norms and easy to operate and maintain.

Some sporadic interventions had been rolled out in the past by different tiers of government. These included the National Urban Sanitation Policy (NUSP) in 2008, which sparked interest in studying the scale of the problem and understanding the requirements. The Odisha Urban Sanitation Strategy (OUSS), 2011, mostly focused on solid waste management with a few remarks on liquid waste management. By 2014, the underground sewerage system was under construction in Puri, Bhubaneswar and Cuttack, at varying rates of progress. Puri was at an advanced stage, being a smaller city, but the Bhubaneswar project was almost abandoned midway by the executing agency and the Cuttack project was also struggling to make progress.

Several significant studies were initiated in the 2010s. An action research study undertaken by Institute for Financial Management and Research in 2011, in partnership with the Bhubaneswar Municipal Corporation, sought to understand and improve the user experience and safety of conventional municipal public toilets. This study was based on the hypothesis that public toilets are not used because of poor design that leads to poor user experience and maintenance difficulties. Another study of the urban sanitation landscape by Practical Action in 2013 underscored the need to promote awareness and understanding among urban households on safe sanitation and hygiene practices alongside infrastructure creation.

As part of a programme called the Scaling City Institutions for India (SCI-FI) initiated in 2013, an analysis based on the results of the 2011 Census was undertaken to identify particular states and geographies where alternative sanitation arrangements such as FSSM and/or low-bore sewer networks could help. This analysis determined that given the low penetration of conventional sewerage systems in Odisha, the cities in the state could benefit from alternative environmental sanitation arrangements. The SCI-FI programme simultaneously undertook detailed analysis of sanitation conditions based on a geographical information system mapping of sanitation facilities in Balasore, a city in Odisha. There was in-depth discussion of these studies with the Housing and Urban Development Department (HUDD) and of research undertaken at SCI-FI on alternative arrangements in countries such as Indonesia, Malaysia, Brazil and Vietnam.

In 2014, Bill & Melinda Gates Foundation (BMGF) organised a scientific and trade exhibition called Reinvent the Toilet in New Delhi. At the workshop organised by SCI-FI BMGF and SCI-FI were asked to pilot a possible alternative solution in Odisha; the result was Project Nirmal. The year 2014 also saw the launch of the Swachh Bharat Mission (SBM) by the central government, with an open defaecation-free (ODF) urban India by October 2019 as one of its key goals.

In 2015, the state set in motion a pilot FSSM initiative in Dhenkanal and Angul under Project Nirmal, with the support of organisations such as SCI-FI, Practical Action, BMGF and the Department for International Development of the UK government. The model to be adapted for Odisha's context was taken from a system put in place in Malaysia. That country has adopted decentralised systems for faecal sludge and septage (FS&S) treatment ranging from deep row entrenchments (DREs) for peri-urban areas to a decentralised wastewater system (DEWATS) with mechanical dewatering and sludge-drying beds for dense core urban areas. FSSM is highly regulated in Malaysia and the services are accessible to the citizens through a digital platform. A modified form of this model was pitched by the Secretary of the state's HUDD in 2016, with a vision of starting the FSSM process in all urban areas of the state.

The Odisha government, with support from SCI-FI, revised the 2009 policy in 2016 to align it more strongly to FSSM. The Odisha Urban Sanitation Policy (OUSP) and the OUSS notified in early 2017 were the first state policies to recognise the need for FSSM, moving away from a sole focus on underground sewerage. SCI-FI's early studies contributed to the formulation of the Odisha sanitation policy and strategy. Later in the year the Government of India (GoI) also came up with the National Policy on Faecal Sludge and Septage Management, 2017, which was close to the approach adopted by Odisha. In another significant development the GoI approved FSSM systems as eligible infrastructure components under the flagship programme Atal Mission for Rejuvenation and Urban Transformation (AMRUT). This was the first occasion when the GoI acknowledged and approved sanitation arrangements other than underground sewerage in its policies and programmes.

As part of the OUSP and OUSS a river basin pollution abatement strategy was also incorporated. This was used as the basis to ensure that 30 cities in total do not dump any untreated sewage into their rivers. The HUDD simultaneously tapped funds from AMRUT to build faecal sludge treatment plants (FSTPs) in the cities identified under the mission.

In 2018, the government undertook the construction of FSTPs in AMRUT cities such as Bhubaneswar, Berhampur, Sambalpur and Rourkela as well as non-AMRUT towns such as Dhenkanal. Odisha's one of a kind Model Faecal Sludge & Septage Management (FSSM) Regulation was also introduced later in the year. For more effective implementation the Odisha government adopted the strategy of convergence of different programmes such as SBM, National Urban Health Mission and National Urban Livelihoods Mission (NULM) by strengthening community-based organisations. In the same year, public–private partnership models were initiated in Dhenkanal along with streamlining operations of cesspool-emptier vehicles. Subsequently, in 2019, an integrated operation and maintenance (O&M) model for Dhenkanal FSTP was implemented with a private operator, Angul FSTP was constructed and an FSSM business model was developed with support from SCI-FI.

After Dhenkanal FSTP was stabilised and was providing city wide services, it was conceived that whether the wider urban region including the villages in the periphery could be served by Dhenkanal city. This led to the concept of urban–rural (UR) convergence in Dhenkanal. In 2020, to build a plug-in model for urban–rural convergence (URC), extensive data collection and analysis was undertaken to guide the planning and implementation processes. During the first phase of the project, 17 gram panchayats (GPs) covering 92 villages within

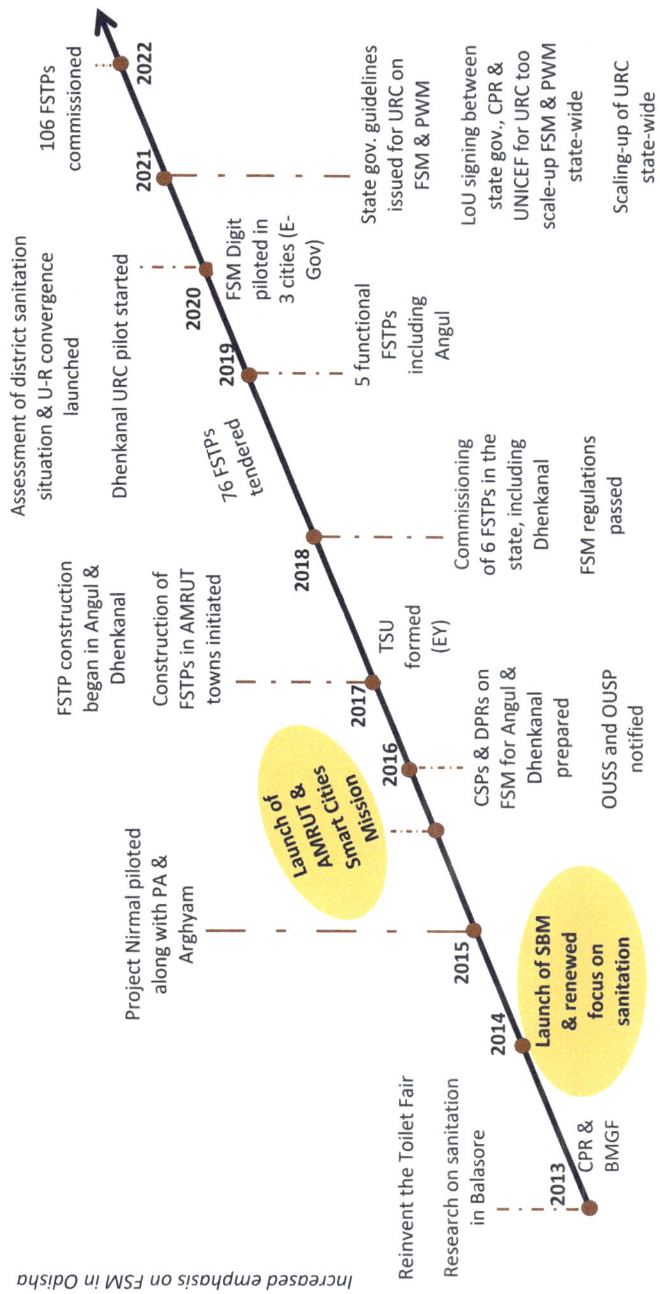

Figure 4.2 Odisha's FSSM journey. *Source:* SCI-FI analysis.

a 10 km radius of the urban FSTP in Dhenkanal municipality were spatially demarcated for the plug-in model, based on the spare capacity of treatment infrastructure and feasibility and cost analysis to ensure sustained access to the existing FSSM facilities (FSTP and cesspool trucks). In parallel, a sample survey was undertaken to identify types of on-site systems in the tagged GPs to estimate the daily faecal sludge loading from the same. Plug-in clusters of 17 GPs were ratified in consultation with the rural and urban local bodies for defining terms of agreement – including desludging tariff, mechanisms for payment and financing and monitoring. Based on consultations, the municipality and tagged GPs passed resolutions to have a Memorandum of Agreement which was finally signed by the respective parties. On this occasion, the district also launched Information Education and Communication (IEC) materials for FSSM in rural areas. Following the agreement signing, the municipality amended the O&M contract of the self-help groups (SHGs) responsible for managing the urban FSTP and desludging operations to extend these services to the tagged rural areas. Finally, extensive capacity building workshops for all the tagged GPs were conducted – covering 670 representatives. IEC campaigns were also undertaken, targeted at rural households, using pamphlets, posters, wall paintings and folk media.

The Dhenkanal municipality provides FSSM services to every rural household in the 110 plugged-in GPs. Data on project implementation is regularly monitored and integrated within the municipality's record keeping system. From February 2021 until February 2022, the urban FSTP has treated 630 kL of sludge from the rural areas. Additionally, 93 GPs in the district have been plugged-in with upcoming FSTPs in urban areas. Furthermore, the success of the pilot project has resulted in agreements between the state government (HUDD and Panchayati Raj & Drinking Water) departments and SCI-FI Centre for Policy Research and United Nation's International Children's Fund (UNICEF) for scaling up the URC model for FSSM and plastic waste management (PWM) in the entire state as shown in Figure 4.2.

4.3 APPROACH AND COVERAGE

Similar to many other states in India, Odisha, too, did not have an underground sewerage system in its urban areas. The dependence on OSS systems was higher still in small- and medium-sized towns (defined as those with a population of less than 100 000), at 88% among toilet-owning households (Figure 4.3).

To provide safely managed sanitation in a sustainable manner in all its cities and towns, the HUDD decided to adopt FSSM as an alternative to sewerage systems at a meeting on 'Improving Urban Sanitation through Septage Management', held on 31 March 2016 under the chairmanship of Chief Secretary, Odisha. Additionally, the state has chosen to hand over the day-to-day O&M of FSPTs to women SHGs created under the state's Mission Shakti programme (led by the Department of Mission Shakti) and transgender groups.

Importance of involving SHGs as stated by Former Collector and Former Municipal Commissioner

Urbanisation is not a quick process; it materialises with infrastructure and, importantly, behavioural change among the people to adapt to such changes. This is why women SHGs were taken on board, signing a tripartite agreement involving the SHGs and cesspool workers. With every order the cesspool operator receives, the SHGs earn a commission of 20%, incentivising them to work hard.

The entire value chain is looked after by the SHGs, from creating awareness amongst the people, construction and maintenance of community toilets to collection and treatment of faecal sludge.

Chakravarty Singh Rathore

(*Source*: Interview, 2022)

The key challenges faced in making FSSM into a robust and efficient functional programme was the site selection for FSTP since there was resistance from the local community. This led to the formation of SHGs to generate massive awareness amongst the people. Through continuous training modules and capacity building interventions they gained confidence in implementing the programme.

The initiative of Bada Khana played a crucial role in capacity building of the SHG workers as it gave them room for open debates. The engagement of SHGs turned out to be a cornerstone of success for implementing FSSM in Odisha.

Sudarshan Chakravarty

(*Source*: Interview, 2022)

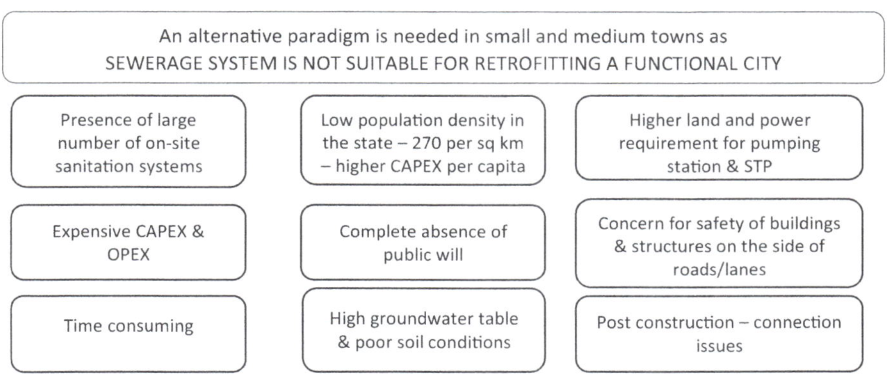

Figure 4.3 Alternate paradigm. *Source*: SCI-FI analysis.

4.4 PROJECT COMPONENTS

Individual OSS systems collect, store and treat (through the natural digestion process in varying degrees) the incoming sewage. Filling up over time, they require periodic emptying. As the evacuated FS&S are only partially treated in the septic tanks and single pits (the two most common types of OSS systems), treating them further before final disposal is a must. FSSM refers to the process of safe emptying, transportation to a stand-alone treatment system, treatment and disposal or reuse of FS&S from OSS systems (Figure 4.4). A typical FSSM system comprises mechanised desludging (or emptying) of OSS systems using a vacuum truck, transportation of the collected FS&S to a treatment facility, its treatment and reuse or disposal of the treated wastewater and recovered biosolids. Here, the treatment facility could be either a dedicated FSTP or a co-treatment-enabled STP.

4.4.1 Enabling components for FSSM adopted in Odisha

In addition to the creation of infrastructure and assets for FSSM, HUDD's initiative on FSSM has been accompanied by a host of interventions focused on people and processes. Targeted at enabling a holistic and behavioural shift to safely managed sanitation (Figure 4.5), these include:

- Simplifying, streamlining and de-technologifying the treatment process and systems.
- **Creating demand** for FSSM through a mix of informational educational communication and interpersonal communication media.
- **Building capacities** of state and local governments, private sector stakeholders and community-based organisations.
- **Addressing regulatory gaps** at the urban local body (ULB) level by formulating and instituting FSSM regulations.
- **Developing robust monitoring and evaluation systems**, using the Sujog platform which integrates both SBM and FSSM programme outcomes.
- **Strengthening supply chains** underpinning FSSM through various entrepreneurship programmes and leveraging the strengths of SHGs in the state.
- **Standardising processes** for improved service delivery through issuing of standard operating procedures (SOPs).

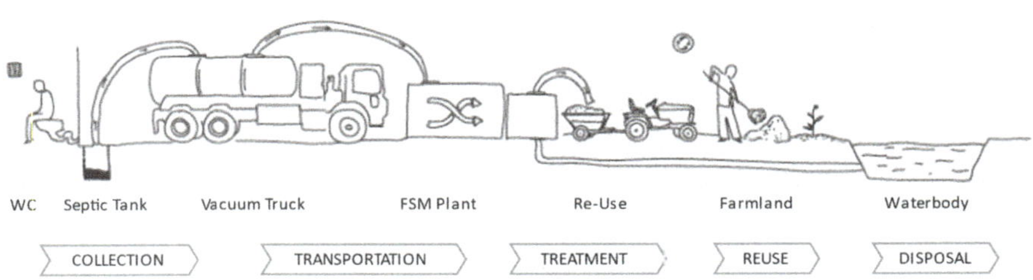

Figure 4.4 Components in FSSM service chain. *Source*: SCI-FI analysis.

Figure 4.5 Aspects of FSSM initiatives in urban Odisha. *Source*: authors' analysis.

- Mainstreaming the issue of financial sustainability for seamless FSSM service delivery.
- Safety and protection for sanitation workers engaged in FSSM.
- Decentralised community-led O&M of the FSTPs.

4.5 INSTITUTIONAL FRAMEWORK

Given the novelty of FSSM as a mainstream approach to wastewater management, the state had to address the lack of a requisite enabling policy and regulatory environment to support implementation. The Odisha Urban Septage Management Guidelines, 2015, constituted the state's first effort in this direction aligned to the OUSS, 2011, the Prohibition of Employment as Manual Scavengers and their Rehabilitation Act, 2013 and the Advisory Note on Septage Management in Urban India, 2013, developed by the Ministry of Urban Development.

In response to the subsequent changes in the landscape of urban sanitation in the state, the government notified the OUSP[1] and a revised OUSS in 2017 (Figure 4), as noted earlier. With the launch of the OUSP and OUSS which provide a comprehensive policy framework aligned to the national environment, public health and occupational safety laws, Odisha ensured its position as a leading light in urban FSSM among its peers. Around this time, the national government too recognised the need to further mainstream and systematise FSSM in urban India, instituting the National Policy on Faecal Sludge and Septage Management, 2017.

[1] Available at https://odxpress.com/wp-content/uploads/Odisha-Urban-Sanitations-Policy-2017.pdf.

As Odisha began to implement urban FSSM at scale under the renewed mandate of the OUSP and OUSS and a newfound impetus from the national policy, it also issued the Model Faecal Sludge and Septage Management Regulations in 2018 to set up a regulatory framework for upcoming citywide FSSM systems. The following sections provide an overview of these instruments.

4.5.1 Odisha Urban Sanitation Policy

OUSP[2] envisions making all cities and towns in the state totally clean, sanitised, safe, healthy and liveable, managed by ULBs and with active citizen and stakeholder participation. The policy is aligned with national policies (National Urban Sanitation Policy, 2008, and National Policy on Faecal Sludge and Septage Management, 2017). Guided by a set of seven principles, the OUSP sets out to achieve six outcomes over a 10-year period (2017–2027). Of the six outcomes shown in Figure 4.6, four are directly related to sanitation and FSSM[3] (Figure 4.6).

4.5.2 Odisha Urban Sanitation Strategy (OUSS), 2017

To facilitate the implementation of OUSP, the HUDD developed the OUSS 2017, which details (a) the institutional framework at the state, district and city levels; (b) provisions and guidance for planning, monitoring and evaluation (M&E), capacity building and training and (c) guidelines for phasing and funding. Based on assessments of institutional readiness for FSSM undertaken as part of Project Nirmal,[4] the OUSS is firmly rooted in the ground realities of the state's specificities. Highlighting the importance of effective coordination and collaboration among the multiple institutions involved in sanitation service delivery across governance scales, it recommends strengthening the institutional structure for effective implementation of FSSM in the state.

Figure 4.6 Planned FSSM outcomes for urban Odisha during 2017–2027. *Source*: OUSP, 2017.

---

[2] Notified vide Notification No. 30586, dated 30 December 2016.

[3] While the other outcomes are also related to sanitation, they do not have a direct bearing on FSSM.

[4] Project Nirmal, operational during 2015–2020, aimed at the demonstration of low-cost, decentralised, inclusive and sustainable citywide FSSM systems for the small towns of Angul and Dhenkanal. A first-of-its-kind initiative focused on sanitation service delivery in small towns, it was implemented by SCI-FI and Practical Action with support from BMGF, Arghyam, HUDD and the municipalities of Angul and Dhenkanal.

4.5.2.1 *Institutional actors*

In Odisha, a variety of institutions are involved in the sanitation sector. Paramount among them is the HUDD which is responsible for 'ensuring proper and planned growth of cities and towns with adequate infrastructure, amenities and services provided to the citizens through ULBs and parastatal agencies'.[5] The HUDD is responsible for implementing the OUSS. It functions through three directorates – the Directorate of Municipal Administration (DMA), Directorate of Town Planning and Public Health Engineering Organization and Water Corporation of Odisha – all of which are actively involved in the sanitation sector. The HUDD also has, under its administrative control, a host of subsidiary organisations[6] through which it seeks to achieve its mandate; key among these in the sanitation sector are the Odisha Water Supply and Sewerage Board[7] (OWSSB), State Urban Development Agency and Odisha Urban Infrastructure Development Fund. Another state institution prominent in the sanitation sector is the Odisha State Pollution Control Board (OSPCB),[8] which is responsible for ensuring compliance with the national environmental laws.[9]

At the local level, ULBs are responsible for implementing sanitation-related initiatives. All ULBs in Odisha have passed FSSM regulations in line with the Model Faecal Sludge and Septage Management Regulations, 2018, developed by the state government. ULBs are also responsible for providing desludging services for OSS systems (either on their own or through private service providers), undertaking O&M of FSTPs, implementation of pollution abatement schemes under the National River Conservation Plan, and execution of sanitation- and FSSM-related initiatives for the workers.

4.5.2.2 *Multi-tiered coordination mechanisms*

As per the provisions of the OUSS 2017, the institutional structure has three tiers with a High-Powered Committee (HPC) as the apex body at the state level. Chaired by the Chief Secretary, Government of Odisha, the HPC is responsible for providing overall guidance and direction to urban sanitation initiatives in the state and overseeing the implementation of the OUSP 2017. The HPC is also responsible for 'ensuring convergence of action for urban sanitation in the State, and to bring about inter-departmental coordination and cooperation'. The HPC

---

[5] *Source*: http://www.urbanodisha.gov.in/About.aspx.

[6] Including Odisha State Housing Board (OSHB), State Urban Development Agency (SUDA), Odisha Urban Infrastructure Development Fund (OUIDF), Valuation Organization and NIHM.

[7] The Odisha Water Supply & Sewerage Board (OWSSB) was established in 1991 as per the Odisha Water Supply & Sewerage Board Act, 1991, for the rapid development and proper regulation of water supply and sewerage services in the state.

[8] The mission is being implemented in nine cities and towns of Odisha: Bhubaneswar, Cuttack, Berhampur, Sambalpur, Rourkela, Puri, Balasore, Baripada and Bhadrak. *Source*: http://www.urbanodisha.gov.in/Amrut.aspx.

[9] Including the Environment (Protection) Act, 1986, the Water (Prevention and Control of Pollution) Act, 1974 and the Solid Waste Management (SWM) Rules, 2016, under the Environment (Protection) Act, 1986.

is supported in the implementation of the OUSP by the State Mission Directorate (SMD), a core HUDD line agency set up for the purpose and to undertake periodic reviews of progress across all line agencies. The Director, SMD, is supported by a Project Management Unit which has sectoral experts (covering the verticals of environment science, project management, IEC and media, information technology and M&E) and support staff on an outsourced basis (Figure 4.7).

Given that most ULBs, especially the small- and medium-sized ULBs, had considerable deficits in terms of capacities to plan, implement and monitor urban sanitation programmes, a district-level mechanism was an appropriate institution for guiding and assisting ULBs on urban sanitation, until such time that ULBs build their own capacities. Thus, as per the provisions of the OUSP 2017, the District Collector (or an officer nominated by the District Collector) has been designated as the nodal officer for urban sanitation-related programmes. Furthermore, a District Urban Sanitation Committee (DUSC) has been created at the District Urban Development Agency as the district-level monitoring and implementing agencies for all urban sanitation programmes, schemes and strategies. Chaired by the District Collector, the DUSC has as its members Mayors/Chairpersons and Commissioners/Executive Officers (EOs) of ULBs, heads of line departments and representatives from local industries' associations, Non-Governmental Organisations and so on. The DUSC remains in constant contact with the SMD (at the state level) to ensure effective coordination and flow of information between all ULBs in a particular district and the state government. The DUSC also assists ULBs in carrying out their tasks as outlined by the City Sanitation Task Force (CSTF). In addition, the district-level Review & Monitoring Committee, under the chairpersonship of a Member of Parliament, ensures satisfactory monitoring of projects.

At the city/town level, a CSTF is created in all ULBs under the leadership of the Commissioner/EO. The CSTF has been entrusted with the responsibility of

Level	Institution
State	**High Powered Committee (HPC)** Chair: Chief Secretary, Odisha; Member-Convener Secretary-cum-Commissioner, H&UDD
	State Mission Directorate (SMD) Mission Director supported by Project Management Unit (PMU), reporting to Commissioner-cum-Secretary, H&UDD
District	**District-level Review & Monitoring Committee (DLMRC)** Chaired by Member of Parliament
	District Urban Sanitation Committee (DUSC) Chaired by District Collector; supported by Programme Implementation Unit (PIU)
City / Town	**City Sanitation Task Force (CSTF)** headed by Commissioner/Executive Officer
	City Mission Directorate (CMD) – headed by Commissioner/Executive Officer; supported by PIU

Figure 4.7 Institutional mechanisms for coordination and collaboration among different agencies at the state, district and local levels. *Source*: OUSS, 2017.

designing, implementing and monitoring all sanitation-related programmes for its city/town. The CSTF is assisted by a City Project Implementation Unit (PIU), headed by the Commissioner/EO. The PIU is responsible for conducting baseline surveys, providing inputs on the City Sanitation Plan (CSP), conducting city-wide communication campaigns, ensuring citizen engagement for sanitation service delivery and developing detailed project reports (DPRs) for urban sanitation under various national and state government programmes/schemes. At the sub-city level, the OUSS recommends the setting up of ward committees or area committees for oversight and implementation of the CSP in coordination with the CSTF.

4.5.3 Model Faecal Sludge and Septage Management (FSSM) Regulations, 2018

Section 388(8) read with sections 390 and 392 of the Odisha Municipalities Act, 1950, empowers Municipalities and Notified Area Councils, to issue regulations, while sections 657, 658 and 659 of the Odisha Municipal Corporation Act, 2003, extend that power to municipal corporations. In line with these provisions, all ULBs in the state have promulgated FSSM regulations in line with the Model Faecal Sludge and Septage Management Regulations, 2018.

The provisions of the Model FSSM Regulations, 2018, cover the entire sanitation value chain and detailed out roles and responsibilities of all concerned actors including owners of premises, cesspool operators, treatment facility operators and the ULB. The details of the regulations are presented in Table 4.1.

4.6 PROJECT IMPLEMENTATION

Prior to the HUDD's systematic intervention to improve urban FSSM state-wide, practices such as unregulated desludging operations by the informal sector, the manual emptying of OSS systems using buckets and indiscriminate disposal of FS&S in the environment were not uncommon. Therefore, the state's first intervention was to create infrastructure that enabled ULBs to deliver reliable and safer desludging and treatment operations. The HUDD was successful in its early experiments in FSSM conducted in the small towns of Angul and Dhenkanal and in nine cities implementing AMRUT. Following this, the department set out to scale up FSSM systems across all 115 ULBs with an adequate fleet of desludging vacuum trucks and (at least) one FSTP per ULB by March 2022.

The state government has leveraged multiple existing programmes and resources for implementation of FSSM in the state. AMRUT funds were tapped for capex for FSTPs, and state finances were utilised for procurement of trucks (NULM-aided funds are not used for IEC) and awareness generation programmes. Rath yatra,[10] PEETHA[11] and Swachh Survekshan[12] are the

[10] Ratha yatra is a Hindu festival associated with Lord Jagannath held at Puri in Odisha. Awareness camps are organised during the festival.

[11] PEETHA is an initiative scheme for spreading awareness of the various government schemes of the state. Awareness camps are scheduled at the GP level on the 15th–20th of each month.

[12] Swachh Survekshan is an annual survey of cleanliness, hygiene and sanitation in villages, cities and towns across India.

Table 4.1 Provisions of the Faecal Sludge and Septage Management Regulation, 2018.

Containment and desludging	Design and construction of containment structures The design, construction and installation of containment structures are to be in accordance with the provisions of the Odisha Urban Septage Management Guidelines, 2016 (OUSMG 2016) *or* any other accepted sound engineering practice issued by the ULB (with prior approval of the state government) or the state or central government (reg 4 (1)). Responsibility of the owner of premises where the toilet is connected to septic tank: • To undertake upkeep, maintenance and safe disposal of sludge from septic tank (reg 4(2)). • To undertake regular desludging of septic tank (every 3 years) or pit latrine (every 5 years) (reg 5 (2) read with Schedule I) – on payment of user fee determined by the ULB which shall be acknowledged through a receipt (reg 4 (3)) – through a septage transport vehicle (STV) registered with the ULB (reg 5 (1)). • To ensure that no pollution is caused due to direct discharge of sludge into open areas/drains due to the malfunctioning or faulty construction of the containment unit (reg 4 (4)). • To ensure mechanical cleaning of septic tanks by ULB or registered cesspool operators with adequate safety measures and no manual cleaning (reg 4 (5)).
Desludging and transportation	Registration of septage transport vehicle (STV) and its operation • Any person, including the ULB, having a cesspool-emptying vehicle shall apply for its registration as an STV to the ULB after fixing GPS with tracking facility and sharing the tracking link with the ULB and on payment of a registration fee (reg 6(1) (2) (3)). • The ULB shall inspect the STV annually for proper functioning of all equipment and compliance with the Motor Vehicles Act, 1988 (reg 6(4)). • The STV shall be registered with the ULB and possess a valid inspection certificate (reg 8). • The registration may be temporarily suspended if the STV does not have the inspection certificate or if it is deemed unfit or if its operation is likely to have negative environmental and health impacts in the opinion of the competent authority (reg 9(1)). • The registration may be cancelled if the owner violates the regulations/terms and conditions of registration or the STV is found not fit to be in service (reg 9(2)). Responsibilities of STV operator during desludging • To carry out desludging using safe and appropriate technology, equipment, protective gear and sound operating practices, which follow the relevant law *or* as per the OUSMG 2016 (reg 12 (1)).

(*Continued*)

Table 4.1 Provisions of the Faecal Sludge and Septage Management Regulation, 2018 (*Continued*).

	• Ensuring that all workers are trained to use protective gear and follow hygiene practices (reg 12 (2) (a)). • To ensure that all safety equipment is operational and in good condition for use (reg 12 (2) (b)). • To ensure that first aid kit, gas detection lamp, fire extinguisher and sufficient disinfectant are kept in the vehicle before proceeding to a collection site (reg 12 (2) (c and d)). • To ensure that sanitary workers do not enter a septic tank under any circumstances (reg 12(2) (e)). • In the event of accidental spillage of sludge or septage, the STV operator shall immediately act to prevent further spillage, minimise the environmental impact and undertake clean-up action, and disinfect the area of spillage by an appropriate method (reg 13(1) (a and b)).
Treatment	Treatment mechanism • The ULB has been given the discretion to approve and notify the manner in which septage is to be processed and treated at the treatment facilities. However, the selected treatment mechanisms have to be in accordance with relevant laws, notifications or as provided in the OUSMG 2016, or any good engineering practices (reg 14 (1)). • Septage or sludge is not to be treated or disposed of in any other manner except as specified and approved by the ULB (reg 14(2)). Responsibilities of the operator of treatment facilities • To ensure adequate safety measures are in place to protect the workers and neighbourhood, proper sanitation facilities in the premises and the disposal of treated septage follows standards laid down by the OSPCB (reg 15 (1) and (2)).
Disposal and reuse	• The treated septage is to be disposed only at a specified location authorised and notified by the ULB which is to be chosen in compliance with the Environment Protection Act, 1986, and the Water (Prevention and Control of Pollution) Act, 1974, and the rules framed thereunder (reg 16(1)). • The location of the disposal site is to be finalised by the ULB in consultation with the OSPCB and the district administration (reg 16(2)). • The treated septage *may* be reused in accordance with applicable rules and safety standards (reg 16(3)).
Powers of the ULB	The ULB is empowered to • levy user charges for providing services with respect to containment, conveyance or treatment (reg 17(2)). • Inspect any premises, STV and septage treatment plant/facility, at any time (reg 18). • Issue directives to remove difficulties in operation of the regulations (reg 22).

(*Continued*)

Table 4.1 Provisions of the Faecal Sludge and Septage Management Regulation, 2018 (*Continued*).

Penalties and fines for violations	• In the first instance, a notice for compliance will be issued to the violator. • In case of non-compliance, a fine of Rs 50 in case of first and second instances of breach, and seizure of property/STV, as the case may be, and a fine of Rs 15 per day in case of third and subsequent instances of breach (Schedule II) shall be imposed on the following persons for the following violations: – Owner of property – unscientific design and construction of containment unit; direct discharge of sewage into drain or road or open area. – Owner of STV – plying of STV without registration/valid certification; failure to attend to accidental spillage. – Owner of treatment plant – discharge of untreated septage. • For other violations of the regulations, a fine of Rs 50 in the first instance and Rs 15 per day in case of continuing contravention shall be imposed.

Source: Faecal Sludge and Septage Management Regulations, 2018.

IEC mainstays in state-wide implementation. Guided by a strong policy and regulatory framework, the state government and ULBs have fostered innovations along the entire sanitation service chain. The following sections discuss the key pillars of the state's FSSM model (except for sanitation workers' safety which is covered in detail in Chapter 9).

4.6.1 Implementing partners

A host of organisations have brought their expertise to bear on urban Odisha's FSSM journey. These include Ernst & Young which forms HUDD's technical support unit (TSU) for urban FSSM; SCI-FI which supported the piloting exercise (with Practical Action), helped with specific issues including the OUSP and OUSS and capacity building and is further contributing by helping the HUDD and Panchayati Raj Department to come up with new institutional models for URC in FSSM (as discussed below) and the Urban Management Centre and WASH Institute which have supported the HUDD's initiatives relating to sanitation worker safety and sanitation training programmes, respectively. Many of these supporting organisations were funded by organisations such as BMGF, Arghyam and UNICEF.

4.6.2 Desludging and conveyance

In 2016, the HUDD made investments to procure cesspool-emptier vehicles to ensure safe emptying and transportation of the contained faecal waste for 57 ULBs of the state's 115 ULBs. These vehicles are equipped with technological options such as global positioning system (GPS) trackers to enable tracking and improvement in service delivery. They also come with personal protective equipment (PPE) to comply with safety norms related to cesspool emptying and transport, and to facilitate mechanised emptying as against manual emptying.

To standardise the cesspool vehicle machinery and specifications, and to ensure that smaller ULBs gain access to the cesspool vehicles, the state procured 86 cesspool vehicles in bulk which were then either operated by the ULBs or given on license to private operators for undertaking desludging operations. The bulk procurement was a one-time exercise which resulted in standardisation and benchmarking of the procurement cost. Furthermore, SOPs for cesspool operations at the ULBs were notified for ensuring safe and scientific desludging.

At first, standardised 3 kL capacity vehicles were procured by the state. Thereafter, smaller vehicles were additionally procured to ensure that the cesspool operators can traverse narrow lanes and provide desludging services without a hassle. Similarly, vehicles are also equipped with long hose pipes to solve access-related issues during desludging. Moreover, many specifications such as the type of pump used in the cesspool vehicles were calibrated by talking with the manufacturers to ensure efficient desludging operations.

Currently, desludging activity is carried out in the state by both ULB-operated and ULB-contracted cesspool-emptier vehicles (operated by private players or SHGs).

In the FSSM value chain, emptying and transportation of faecal sludge/septage is a crucial component. To provide efficient services to citizens, avoid indiscriminate disposal of the waste and streamline services, the government has taken various measures as follows:

- Issuing guidelines for GPS monitoring of cesspool-emptier vehicle fleets.
- Wearing PPE during service delivery.
- Opening of an escrow account for maintaining transparency of financial transactions between ULB and outsourced operator.
- Mandatory disposal either in DREs as an interim measure or at the FSTPs.

At the ULB level, steps to strengthen cesspool-emptier vehicle operations have been taken; these include rationalisation of prices, setting up cesspool-emptying ULB-level service requisition cells, appointment of a nodal officer for FSSM, training of cesspool-emptier operators and disposal at designated place/s.

Registration of all cesspool vehicles operating in the urban areas was mandated under the Model FSSM Regulations passed in 2018, and the cesspool vehicle operators were invited to register with the concerned ULB. This rule was established to track the cesspool vehicles and prevent indiscriminate disposal of sludge. Penalties are imposed for dumping faecal sludge in the open ground including fines and seizing of the cesspool vehicles to dissuade private operators from discharging in undesignated places. However, this turned out to be a long-drawn process, especially in bigger corporations, owing to the large number of private operators plying in the bigger ULBs along with difficulties in persuading unwilling private operators to get registered. Therefore, a series of consultations were held with the private operators, along with arrangement of exposure visits to the FSTPs, to bring the private players on board. Moreover, as it was difficult to track and monitor indiscriminate disposal in bigger cities and towns due to the size and scale of operations, city-level enforcement committees were set up in municipal corporations comprising representatives from the police, transport and municipal administrations. These committees are responsible for monitoring, reporting and acting against indiscriminate disposal of sludge. Due to the involvement of various actors in providing desludging service to the citizens, after successful piloting of a digital platform in three towns: Berhampur, Balasore and Dhenkanal, it has been scaled up to 69 towns, 33 in the first phase and 33 in the second phase excluding the initial three pilot towns. This platform serves as a single window for the citizens to avail the service.

For the cesspool vehicles owned and operated by the ULB, a cap of Rs 500 was set per trip. Similarly, for the vehicles owned by the ULB but operated by private service providers, a cap of Rs 1000 per trip was set for 3 kL capacity vehicles and Rs 700 per trip for 1 kL capacity vehicles. The main aim of this exercise of fixing price caps was to ensure affordability of desludging services to incentivise the uptake of the service. Another advantage of rationalising price is that it helps in maintaining a fair and uniform price for cesspool-emptying services for both the ULB-operated vehicles and ULB-owned private-operated vehicles. Training of cesspool-emptier operators and ULB staff helps in understanding the critical issues related to mechanical emptying and associated safety measures. It also helps in improving the process of service delivery, safe disposal and providing solutions for the same.

4.6.3 Treatment

After rolling out the pilot implementation of FSTP in eight cities DRE has been adopted as an interim measure to ensure safe disposal of septage desludged from the septic tanks. To pursue this, OWSSB issued a detailed guideline for safe functioning and operation of DRE, the highlights of which are mentioned below.

- The ULB would identify a site for disposal of septage/faecal sludge adopting the method of DRE and kept the trenches of appropriate length and width ready to receive sludge.
- Vehicles were required to take optimal path considering the importance of lane/bye lane to reach the interim disposal site.
- Park truck close to the trench for disposal but at a safe distance to avoid collapse of soil into trench.
- Discharge the contents into trench with help of hoses.
- Any accidental spillage should be cleaned immediately using proper absorbent material. The collected spillage should be treated as normal septage/faecal sludge.
- Clean cesspool container, hose pipe and other equipment using water available at site and discharge the cleansed wastewater to the DRE.
- Use PPE and follow safety operation procedures during loading and unloading activities.

It was a bold step taken by Odisha to adopt DRE as an interim approach to prevent the contamination of ground and river water sources in the absence of any precedence and guidelines from pollution control boards. However, Odisha Pollution Control Board (ODPCB) was on-boarded to study the environmental impact of the model and approval was obtained. Odisha was the first state in India to strategically adopt DRE as an interim treatment solution strategy for safe disposal of FS&S while FSTPs were being proposed or under construction across ULBs. This proved effective in the absence of functional treatment facilities at the time, and served as part of a graded, step-by-step approach, curbing indiscriminate disposal of sludge. Exposure visits to DRE arrangements in Malaysia helped officials in Odisha to adopt this arrangement by instituting a set of guidelines for identifying areas suitable for DREs. National-level policies have now recognised the value of having DRE arrangements; the Department of Drinking Water and Sanitation, Ministry of Jal Shakti, published an FSSM toolkit for district-level officials in 2021, suggesting the uptake of DREs as an interim arrangement where sewage or faecal sludge treatment facilities are either absent or under construction. All the FSTPs in the state employ the DEWATS and sludge-drying beds as underlying technology. DEWATS has been chosen due to its simple nature-based process design, low requirement of energy, ease of O&M and a consequently overall low cost of treatment. The primary configuration of an FSTP sequentially consists of:

- Inlet chamber with screening: for receiving sludge from vacuum trucks and separating any domestic solid waste products from it before further processing.
- Settling-and-thickening tanks: for fractionation[13] of the septage and faecal sludge into solid and liquid streams.
- Sludge-drying beds: for dewatering of the thickened sludge and drying of the retained solids until the point of refinement.
- Anaerobic baffled reactor and anaerobic filter (ABR-AF): for treating the liquid effluent from settling-and-thickening tanks and sludge-drying beds.
- Planted gravel filter integrated with sand filter bed: for nutrient removal and further remediation of partially treated effluent from ABR-AF.
- Polishing pond: for inactivation of any remaining pathogens in the treated wastewater.

All plants have landscaping as an integral part of the design to increase the community's acceptance of the presence of waste treatment plants in their locality. As FSTP O&M required deployment of new operators and training and capacity building of human resources, the state decided to develop these skills within the community instead of hiring private operators. This step was taken to create employment opportunities for the local community as well as to create service providers locally. SHGs are engaged by the ULBs through a contract agreement signed by the two parties and receive a fixed monthly service charge for undertaking O&M of the FSTPs. A technical resource person from the local community is hired by the SHGs to operate and maintain the FSTPs, paid by and held accountable by the SHG. The technical resource person is a chemistry

[13] Fractionation processes are processes of separation of mixtures of liquids, gases or solids into their components.

graduate trained in laboratory operations who acts as a lab technician; he/she coordinates with the nodal officer at the concerned ULB. Appointing a technical resource person has eliminated the need for appointing an engineer at the FSTP. (The ULB Nodal Officer (ME/JE) continues to monitor.)

The OWSSB prepared DPRs for construction of FSTPs in 115 ULBs. However, some major challenges such as land identification and on-boarding of private contractors slowed down the construction process. To fast-track construction of FSTPs, 30 FSTPs were identified, and city-level teams for FSTP construction were formed under the leadership of the EO of the respective ULBs with the concerned contractors, ULB nodal officer and TSU members. These 30 teams, called the 'SUPER 30' were invited to Bhubaneswar for orientation, agenda and target setting. Senior officials including the Principal Secretary of the department spoke at the meeting to inspire and motivate. Dedicated WhatsApp groups were formed at the city and state levels to flag implementation issues for quick resolutions. Senior officials including the Principal Secretary, DMA and others were made part of the group to intervene instantly whenever needed to resolve the problems. With constant follow-up and a mission mode approach, not only these 30 but another four cities – which got inspired and joined in – completed the construction and commissioning within a record time of just 4 months. Through this experience, a strategy was devised and adopted for speedy construction of FSTPs across the state.

4.7 URC FOR FSSM

Odisha faced an extremely low level of access to a toilet facility among rural households at 13.75% as per Census of India 2011. However, the share steadily climbed up to a purported 100% during 2014–2019 under the SBM Gramin. Most of the old and new toilets alike – a cumulative total of 90% of all toilets as per the National Annual Rural Sanitation Survey 2019–2020 – are associated with septic tanks and single pits. Being endowed with a high groundwater table, the state can only derive negligible benefits from technologies such as twin pits that sanitise stored faecal sludge completely on-site under ideal and drier conditions. As a result, rural Odisha too has found itself contending with the need for FSSM to achieve safely managed sanitation in recent years. Responding to this need for FSSM among rural households while maximising capacity utilisation of urban FSSM systems, ULBs in Odisha began catering neighbouring rural areas as well.

In Dhenkanal and Balasore districts, piloting was performed to extend FSSM services such as cesspool vehicles for septic tank desludging in rural areas and treating the sludge in the FSTPs located in the municipalities. Encouraged by the successful implementation of pilot projects of URC in Dhenkanal[14] and Balasore municipalities, the HUDD and its rural counterpart, the Panchayati

[14] The pilot with Dhenkanal Municipality is part of the 'Dhenkanal District Pilot Project for Solid and Liquid Management' implemented by SCI-FI with support from UNICEF, the Dhenkanal district administration and the Dhenkanal ULB.

Raj and Drinking Water Department (PRDWD), decided to adopt it at the state level. This initiative seeks to (1) leverage the FSTP facility available in a nearby town for rural households instead of creating another FSTP (which may not be financially viable and feasible due to size and scale); and (2) utilise the available/surplus capacity that exists in the FSTP created in the ULB and to avail capex and opex for the same from the rural bodies/PRDWD. A joint letter from the two departments issued in May 2021 laid the groundwork for scaling URC for both FSSM and PWM in the state. Therefore, as more and more FSTPs are being commissioned and operationalised, associated ULBs are identifying the rural settlements within 20 km from the city or town for inclusion in their service area. The letter specifies the distance-based tariff that would apply to rural households for availing of the urban FSSM services to ensure the long-term financial sustainability of the model.

SCI-FI and UNICEF with support from the district administration of Dhenkanal, ULBs and Panchayati Raj Institutions piloted the novel URC approach in Dhenkanal district in October 2019. Under this pilot demonstration, a ‘plug-in model’ for tagging rural GPs with Dhenkanal municipality was built by using a step-by-step approach wherein: (a) spatial analysis was conducted to identify ‘plug-in zones’ (within a 10–20 km radius from urban areas); (b) a sample survey was conducted to estimate the rural faecal sludge loading; (c) the capacity utilisation trend for Dhenkanal FSTP was analysed to estimate spare capacity and (d) stakeholder consultations were organised to build consensus among urban and rural local bodies. Thereafter, a Memorandum of Agreement was signed between Dhenkanal municipality and tagged rural GPs to codify the terms of agreements such as tariff structure for desludging services, mechanisms for payment and financing and monitoring. Additionally, IEC campaigns were launched in the tagged rural GPs to spread awareness and generate demand for FSSM services. Since the launch of the pilot demonstration in Dhenkanal, the urban FSTP has received and treated over 1000 kL of rural sludge from the plugged-in GPs in the district.

4.8 PROGRESS ON THE GROUND AND IMPACT

Beginning with a single STP and only 2% of urban faecal sludge being treated in 2015, the state is amid enabling safely managed sanitation for all its urban population and a significant share of its rural population by 2022. In 2019, 82.5% of faecal sludge from urban areas was already being safely collected, conveyed and treated. By March 2024, a total of 119 FSTPs out of the 120 FSTPs planned are fully operational remaining only one to be completed. Besides setting up infrastructure, the state has also been continuously handing over the operations of the FSTPs to SHGs and running large-scale capacity-building programmes for the SHGs on a regular basis.

O&M has often been the Achilles heel of urban service delivery systems. Creating FSSM systems requires continuous monitoring and improving their operational and financial performance. To this effect, the HUDD has been piloting an open-access digital governance platform for FSSM in three urban local bodies – Balasore, Berhampur and Dhenkanal – since December 2020 which has now scaled up to 69 towns. The platform allows citizens, service providers and the ULB to engage in a trackable and transparent manner. While citizens have an online channel for requesting services, the ULB and service providers can manage the end-to-end logistics of desludging service delivery on a single platform. The platform collates data for all these interactions along the service chain and visualises it in monitoring dashboards. It is expected to enable ULBs to improve service delivery and planning by fostering greater visibility over the FSSM system's performance.

Overall, Odisha has been one of the first states to demonstrate its commitment towards making all urban areas (cities and towns) not just ODF[15] but also open discharge-free. The state has adopted an incremental approach for ULBs to attain an ODF status, progressing from stage 1 to 3, that is, ODF, ODF+ and ODF++.

Successful implementation of FSSM in the state has also brought about tangible pollution abatement in the major rivers flowing through the state. A significant and consistent reduction in total coliforms was observed between 2015 and 2020 in the Mahanadi, Kathajodi, Kuakhai, Daya and Budhabalanga, as reported by the OSPCB; samples were taken from upstream and downstream of the rivers crossing urban centres/districts such as Sambalpur, Cuttack, Bhubaneswar and Balasore where FSSM interventions have been implemented (Figure 4.8). Latest CPCB report on water quality of rivers monitored under national water-quality monitoring programme, 2021 showed reduction in the pollution level in 12 out of 19 river stretches of Odisha.[16] All cities in Odisha today have FSSM infrastructure and systems in place and urban FSSM team is extending the services and mentoring the expansion of the FSSM activities in the rural areas.

4.9 RECOGNITION

The progress made by the state of Odisha in the FSSM sector has been well recognised at the national and international levels. In 2018, a national workshop on FSSM was organised for the first time in India. By that time, some other states had started working on FSSM, but the conference was historic as it was the first national-level event on the subject and in being organised by Odisha, a torchbearer of FSSM in the country. Senior officials at the level of Secretaries of the HUDD and Directors of SBM from states attended the event in Bhubaneswar. This workshop played a significant role in generating greater interest in FSSM. The conference was inaugurated by the former Chief Minister of Odisha and

[15] Open defaecation-free is defined as (a) no observed open defaecation; (b) all city residents have access to and use household, community and/or public latrines; (c) there is adequate access and use of latrines in all institutions; (d) all insanitary latrines (including single-pit latrines) are converted to sanitary latrines, and no incidence of manual scavenging observed and (e) all city residents are engaged in safe hygiene practices, including hand washing.

[16] https://cpcb.nic.in/wqm/2021/WQuality_River-Data-2021.pdf.

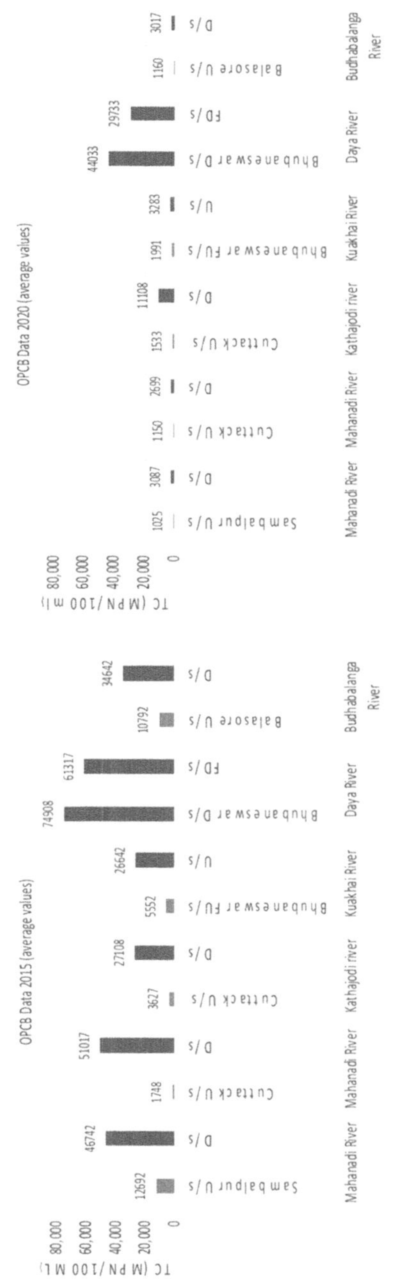

Figure 4.8 Microbial river water quality monitored by SPCB in different towns in 2015 (left) and 2020 (right). *Source*: SPCB, 2015, 2020.

attended by the former Secretary and former Additional Secretary, Ministry of Housing and Urban Affairs, GoI. Following this event, many government and non-governmental representatives, including representatives from countries such as Bangladesh and Nepal, visited Odisha to study and learn FSSM uptake and progress, especially the community engagement model adopted by the state. Exposure visits to Odisha led to the state's experience feeding into the national-level policy ecosystem, such as the DRE interim arrangement featuring as a directive, and engagement of community-based organisations in infrastructure management and model FSSM regulations.

In 2020, Odisha received the ISC-FICCI Sanitation Award at the India Sanitation Conclave for FSSM. All the FSTPs commissioned so far are ISO-certified, with the FSTP in the Berhampur Municipal Corporation being the first FSTP in India to receive ISO certification. Berhampur Municipal Corporation also bagged the Best Faecal Sludge and Septage Management Model Award (Urban) at the national level. The ULB was lauded for its innovative, community-driven and sustainable FSSM. Furthermore, the Agrata City Level Federation of Mission Shakti SHGs in Berhampur was awarded the 'Best Engagement Model in Sanitation' by social enterprise or a start-up. Under the National Urban Livelihood Mission, a group of SHGs federated at the city level constitutes a City-Level Federation (CLF) in cities where the mission is being implemented. In Odisha, the first CLF was instituted in Berhampur, named Agrata; it is one of the most successful CLFs in the state, entrusted with end-to-end responsibility of the entire sanitation value chain. Berhampur had close to 700 Mission Shakti SHGs constituting 7000 women members, and the ULB decided to partner with these SHGs, engaging them across the sanitation value chain – from toilet construction, septic tank desludging and demand generation to O&M of the FSTP.

Moreover, the first six cities of Odisha to acquire an FTSP have been recognised as pioneer cities at the national level by the National Centre for Good Governance, Mussoorie.

Malasur IEC campaign, in collaboration with BBC Media Action, implemented the Malasur campaign in Municipal Corporation, Berhampur, and was awarded with the 'Best Faecal Sludge and Septage Management Model (FSSM)' by ISC-FICCI Sanitation Awards, 2020. The campaign took a unique approach towards underlining the need for FSSM by personifying faecal sludge as a 'demon' and providing information on how this 'demon' could be stopped from entering people's homes to drive the adoption of correct sanitation practices.

At the fifth ISC-FICCI Sanitation Awards in 2021, too, the successful demonstration of the innovative URC model for FSSM won Dhenkanal Municipality the 'Best Faecal Sludge and Septage Management Model (Urban)' award. Bahucharamata transgender SHG was honoured with the 'Women Changemakers in Sanitation' award at the national level by ISC-FICCI for her leadership in managing the FSTP in the city of Cuttack with support from the municipal corporation. Again in 2023 Meghna Sahoo, president of TG Swikriti group managing the FSTP at the Basuaghai was honoured with the 'Women

Changemakers in Sanitation' award at the national level by ISC-FICCI. Again, in 2022, Scaling City Institution for India initiative was felicitated for its URC project in Dhenkanal at the sixth ISC-FICCI Sanitation Awards as the 'Best Non-profit Engagement Model in Sanitation (Urban)', whereas Urban Management Centre was felicitated for 'Best Communication in Sanitation: Digital'.

Overall, Odisha has made its mark in the realm of urban services by exemplifying a low-cost and sustainable approach to sanitation in urbanised and urbanising contexts. It is observed that innovative interventions made in Odisha bring laurels for the state as well as the partners working with the state. The FSSM model of Odisha has not only transformed the black water management in the state but also has created large-scale livelihood opportunities for the marginalised section of the society including women and transgender groups making them as sanitation professionals.

Situation before FSM

Source: HUDD, GoO.

Balasore FSTP

Source: HUDD, GoO.

Cesspool vehicle disposing sludge at the Bhubaneswar (Basuaghai) FSTP

Source: HUDD, GoO.

Cesspool vehicle operation during mechanical desludging in Berhampur

Source: HUDD, GoO.

Berhampur FSTP

Source: HUDD, GoO.

FSTP Bhadrak

Source: HUDD, GoO.

doi: 10.2166/9781789064926_0059

Chapter 5

Municipal solid waste management with community partnership

ABSTRACT

The poor management of municipal solid waste causes diverse environmental and health risks across spatial and temporal scales. This chapter charts the state's trajectory in developing and scaling a unique model for urban solid waste management that valorises decentralisation, local ingenuity, circular economy and community partnerships.

Keywords: Solid waste management, micro-composting centres, material recovery facilities, community partnership, organic manure, women self-help group, transgender group, battery-operated vehicles, recyclable, non-recyclable.

5.1 OUT OF SIGHT, OUT OF MIND

How often do we think of what happens to garbage once it has left our houses? Often, residents of a locality question this issue only if they face inconvenience caused by a dump yard or a landfill or a garbage incinerator not too far away.

Although urban local bodies (ULBs) assign high priority to beautification of the city, not as much effort goes into ensuring that today's waste does not become the gestation point for tomorrow's problem. Traditional systems of dealing with garbage follows a 'truck and dump' approach where waste is collected and merely dumped in open areas without downstream processing and disposal of waste. Solid waste management (SWM) remains a major urban issue around the world, especially in the rapidly urbanising Global South.

Landfills, in one form or the other, have been the approach for managing garbage for millennia. However, with the growing levels of chemicals, toxic substances and non-biodegradable materials in our trash, a wide range of troubles is emerging: foul-smelling neighbourhoods, huge hills of garbage on the outskirts of cities, release of noxious or carcinogenic gases through deliberate or accidental burning of garbage, groundwater contamination and so on.

5.2 QUICK MOVEMENT

Until 2020, Odisha's ULBs did not have an effective system to manage the municipal solid waste and ensure scientific process and safe disposal. In 2018, at least 30% of urban households in the state reported that they disposed their garbage directly into their environment. This mostly meant the trash was thrown on nearby empty land or out on to the street.

The central government's Swachh Bharat Mission envisioned SWM as one of its components. It required ULBs to prepare detailed project reports for citywide SWM systems (SWM-S), and accordingly the central government would bear the cost of this process. The central government was to pitch in a maximum of 35% of the total project cost as a grant or viability gap funding. However, the stronger push under the mission in most states has consistently been for the construction of toilets and making the country open defaecation-free. Odisha, however, prioritised the implementation of SWM since 2019 and is now a front runner in decentralised SWM.

The inspiration and reference point for putting in place a robust and self-contained SWM-S in Odisha was the system followed by a few municipalities in the outskirts of Chennai. In 2019 a team of officials from Odisha visited Chennai to study some functional facilities, including micro-composting centres (MCCs) and material recovery facilities (MRFs). After deliberations on the Chennai model and contextualising it for Odisha, a specific model for Odisha was built incorporating key ingredients of the Chennai model with community partnership as its backbone. While inspired by the Chennai system, Odisha's model has incorporated a wide array of changes to ensure it can be made to work at the ground level at scale – covering the entire urban space administered by all ULBs in the state.

Odisha's SWM journey moved quickly, going from ideation to concept note to making standard operating procedures (SOPs) for training to launch this in a time span of just 3 months, which includes approval of competent authorities at various stages and levels. It was launched on 30 August 2020, the state's Local Self Government Day.

5.3 OVERALL APPROACH

The state of Odisha set out to create an SWM-S that is scalable and, viable across diverse local contexts, and operationally sustainable. Informed by past experiences, the following three concepts underpin the state's approach in the pursuit.

5.3.1 Eliminating 'ghost garbage'

Typically, urban infrastructure programmes in India have promoted public–private partnerships (PPPs) for large-scale undertakings such as highways and smaller urban services projects such as SWM and faecal sludge and septage management.

Three of the oft-cited reasons for encouraging PPPs for such projects are:

- to bolster constrained public resources by inviting private capital for funding infrastructure and services;
- to bring in 'private sector efficiencies and innovation' for better service delivery and its management;
- to supplement limited capacities and competencies of ULBs in delivering projects.

PPP contracts have indexed payments to the private operators to the volume or weight of solid waste required to be processed. As urban services such as waste management are considered 'public goods', the capital-intensive SWM-S typical of PPP projects have limited potential for cost recovery especially in small and medium towns. Therefore, private operators are known to resort to malpractices with the connivance of local officials to increase their revenue. Reporting an amount of garbage that is more than what is being produced and billing ULBs for 'ghost garbage' is one of them. It is not surprising then that earlier studies have established that PPPs, in many cases, have come at a high cost to the public purse, constituting a well-established source of misappropriation of such funds. To avoid siphoning from public funds towards 'ghost garbage' and high-risk PPPs that are complex to develop and manage, Odisha sought to create a model that relied on public sector expertise and accountability (building capacities towards both as required) tailor made to its own requirements. Another disadvantage of engaging private sector for end-to-end management of delivery of critical public services such as sanitation or water is that the government machinery abdicates the entire responsibility to the private partner diluting its own accountability and control which often results in failure of PPP projects.

5.3.2 Low appropriate technology or no technology

The very simple principle of 'My Waste, My Responsibility' has been adopted, with households handing over segregated waste – of two types, that is, dry and wet waste – during door-to-door collection every day. The wet waste is processed in MCCs adopting the process of aerobic decomposition and the dry waste is further segregated into recyclables and non-recyclables; the former is sold, and the latter is sent to cement plants with kilns for use as secondary fuel. Community partners including women's self-help groups (WSHGs), transgender groups (TGs) and waste pickers groups manage the facilities from end-to-end. Care was taken not to employ high-end technology; rather an appropriate technology that is as simple as possible was employed so that the community partners could operate and manage it with ease. The groups are imparted with adequate training to familiarise with the technology, process and the operation and maintenance (O&M) aspects. This enabled an easy and fast rollout of the system across the entire state and leaving ideally no waste to go to landfills as shown in Figure 5.1.

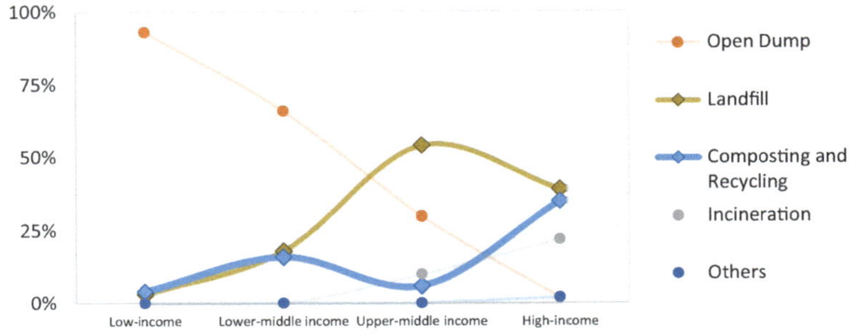

Figure 5.1 Solid waste disposal methods by country income. *Source*: adapted from Kaza *et al.* (2018).

5.3.3 Decentralisation

'Community driven' (a change from the hitherto 'contractor driven' approach) and 'decentralised' (as opposed to 'centralised') are the two core principles adopted for the implementation of SWM in Odisha. The principle of 'decentralisation' is fundamental to the 74th Constitutional Amendment mandating the devolution of powers to wards as the lowest unit of governance in cities and towns. Besides, it encourages community participation in SWM and helps in establishing a 'circular economy', with the 'wealth' generated from waste flowing back to the community partners as monthly incentives.

5.4 INSTITUTIONAL FRAMEWORK

The responsibilities of national, state and local governments in India in terms of SWM are enshrined in the Solid Waste Management Rules, 2016, issued by the Ministry of Environment, Forest & Climate Change. The rules cover ULBs, outgrowths in urban agglomerations, census towns,[1] notified industrial townships and standalone entities such as airports, ports, harbours and others. They stipulate that a state's Urban Development Department Secretary (through the Director of Municipal Administration) would prepare a policy and SWM strategy for the state and ensure that ULBs implement all the provisions of the rules. The actual operationalisation of the strategy – including, inter alia, aspects such as tariff setting, issuing of apposite byelaws, creation of infrastructure and assets, enhancement of public awareness – would rest with the ULB. The State Pollution Control Board is to facilitate and monitor the progress of the state in ensuring solid waste is managed in accordance with the rules.

Following a reinvigoration of Odisha's agenda on SWM in early 2019, the state issued an SOP for decentralised SWM to all ULBs on 30 July 2019. The

[1] In India, a census town is one which is not statutorily notified and administered as a town, but nevertheless whose population has attained urban characteristics.

SOP codified the state's strategy and laid out a comprehensive step-by-step process for ULBs to implement SWM, incorporating:

- quantification of waste generation;
- assessment of the need for human resources;
- engagement of human resources;
- capacity building of human resources through community partnership;
- identification of land for development of infrastructure;
- assessment of requirement of solid waste collection vehicles;
- preparation of route maps;
- source segregation and transportation of waste from source to the processing units;
- method of treatment and processing of solid waste;
- method of channelising disposal of the processed/recyclable for onward use and safe disposal of non-recyclable solid wastes;
- handling of bulk waste generators;
- implementation of information, education and communication campaigns;
- mechanisms for financing;
- timeline for implementation;
- real-time monitoring of waste collection, processing and safe disposal;
- evaluation and performance assessment.

To enable a faster scale-up, alongside the SOP, the state issued model plans (along with estimated costs) for solid waste processing facilities of different capacities (refer to Section 5.5): 5 tonnes per day (TPD), 4 TPD, 3 TPD, 2 TPD and 1 TPD. It also shared sample quotations and specifications for equipment required by the facilities; to help the ULBs procure the necessary equipment through the Government e Marketplace portal, ensuring transparency.

5.5 PROJECT COMPONENTS

An SWM-S typically comprises the sequential chain of collection, transportation, processing and final disposal or recycling of both the biodegradable and non-biodegradable categories of solid waste. Building on its 'low technology' and 'decentralised' approach, the state has designed a simple, community-centric, cluster-based SWM-S to foster a cleaner and healthier environment citywide. Instead of creating centralised infrastructure at the level of the city, the state has opted to implement SWM-S for smaller and more manageable clusters of five to six wards in each ULB. Each SWM-S leverages the strengths of the strong network of Mission Shakti[2] WSHGs, TGs and waste picker groups in

---

[2] 'Mission Shakti' is the self-help mission for empowering women through promotion of WSHGs to take up various socio-economic activities which was launched in the state on 8th March 2001 on the eve of International Women's Day. Mission Shakti has the clear objective of empowering women through gainful activities by providing credit and market linkage. Empowerment of women through WSHGs under Mission Shakti is a flagship programme of the government. It envisages that over a period of time more and more women would be part of a WSHG.

the state as Swachha Sathis and Swachha Karmis to manage the services and the underlying infrastructure (refer to Section 5.6). The following subsections elaborate on the cluster-level SWM-S and its different components.

5.5.1 Collection

Urban Odisha's SWM-S model adopts and promotes source-level, or household-level, segregation of solid waste into its three constituent factions for improved efficiency and sustainability of the overall system. The designated Swachha Sathis from WSHGs go on daily rounds to create awareness and educate the households and bulk waste generators such as restaurants, hostels and so on to hand over segregated waste.

The state has paved the road to source-level segregation with extensive measures to induce household behavioural change. In the beginning, all the 115 ULBs in the state undertook door-to-door awareness and sensitisation campaigns with the help of 28 606 Swachha Sathis and Swachha Karmis. Second, the state employs the '3Es' principle of expose–embarrass–educate to ensure the continued practice of source segregation of solid waste. Applying the 3Es principle, the Swachha Sathis select households at random and check the waste they have handed over during the daily collection rounds to 'expose' cases of improper segregation or non-segregation. By conducting the check publicly, as seen in photos below, the Swachha Sathis can 'embarrass' the households that have not been segregating their waste. With embarrassment acting as a potential deterrent to the household defaulting to non-segregation again, the Swachha Sathis proceed to 'educate' the household on how to segregate and store their waste. Third, the ULB recognises the efforts of households that have demonstrated proper compliance with the practice of source segregation by bestowing a 'Swachha Gruha' sticker on their premises.

5.5.2 Transportation

The Swachha Karmis drive battery-operated vehicles (BoVs) and light commercial vehicles (LCVs) for transportation of the segregated waste from door to door and commercial areas to the cluster-level waste processing facility. All vehicles have separate storage compartments for wet and dry waste to ensure that the collected segregated waste travels segregated to the facility.

There is sector-level route mapping for individual collection vehicles, which aims to minimise the time taken and power/fuel consumed to complete a daily round. This ensures predictability of collection time for the urban residents and systematises the monitoring of source segregation.

The current plan for route mapping envisages the further subdivision of sectors into individual spatial loops. Each loop would encompass an average of 150 premises with the expectation that the Swachha Karmis will cover these premises in one seamless route and within a definite start and end time. Keeping with the spirit of decentralisation, the Swachha Sathis and Swachha Karmis operating in the specific sector would develop their route map with guidance from the respective circle-level supervisor and a sanitation expert posted in the ULB as is shown in Figure 5.2.

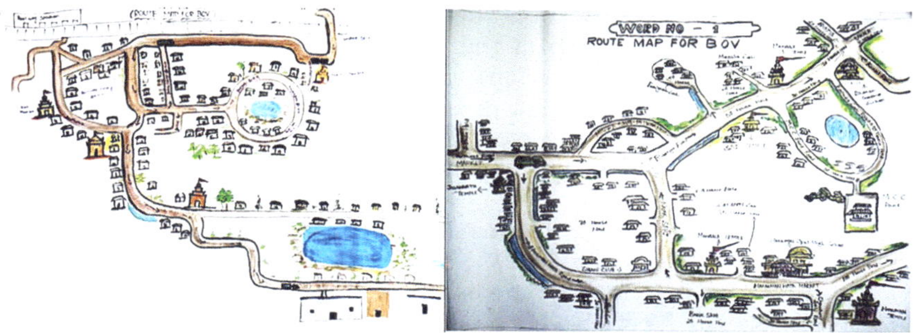

Figure 5.2 Examples of route maps prepared by sector-level team. *Source*: collected from field by SCI-FI team.

Additionally, three interventions aim to facilitate the streamlining of operations and the adoption of optimised routes.

(a) Signage in each lane that specifies the contact information of the relevant Swachha Karmis and the timing of solid waste collection for the residents' general information.
(b) Mapping of all urban residences and their categorisation based on their adoption of source segregation using a three-point scale: green (households providing segregated waste consistently), orange (households providing partly segregated waste or segregated waste inconsistently) and red (households providing non-segregated waste).
(c) Maintenance of daily records by Swachha Karmis and Swachha Sathis on the number of premises covered, total quantity of waste collected and extent of adoption of source segregation.

5.5.3 Processing

Each cluster-level operated SWM facility, called a wealth centre, is designed to process 6–10 TPD, and is run and managed by either a WSHG or TG. It has a 3–5 TPD MCCs for processing the segregated biodegradable (wet waste) and 3–5 TPD MRFs for processing non-biodegradable waste streams (dry waste).

(a) *Biodegradable waste*: At the MCC, the facility operators inspect the biodegradable waste stream and undertake a final round of segregation, if required, to eliminate any mixed-in non-biodegradable elements. The waste stream then passes through a shredder that breaks up bigger pieces of waste into smaller ones for homogeneous processing. In the next step, the shredded waste makes its way to a composting tub, where it is mixed with a locally prepared solution of effective microorganisms.[3] As with all types of composting, the waste mixture is periodically turned over until it decomposes and dries out to the final product: a compost named 'Mo Khata' (My Manure). The compost thus produced passes through a mechanical sieve, which crushes it to make it finer, and is packaged for sale.

(b) *Non-biodegradable waste*: The non-biodegradable waste undergoes sorting as the first step at the MRF. The facility operators sort the incoming waste into recyclable and non-recyclable plastic, glass, metal and so on (compressing low-density plastics into bales) and store these separately for further disposal or recycling.

5.5.4 Disposal or recycle

(a) All the final product streams from MCCs and MRFs contribute to revenue generation for the SWM-S. The compost produced at the MCC as 'Mo Khata' or 'My Manure' is sold at designated 'Mo Khata' outlets.
(b) Similarly, the recyclable parts of the non-biodegradable waste are made available for sale to authorised recyclers. The non-recyclable combustible components find their way to one of the four identified cement factories in the state for use as secondary fuel.

5.6 PROJECT IMPLEMENTATION

With the enduring effectiveness of the SWM-S as the goal, the state has created an informed and innovative implementation framework, foregrounded in sustainability, for scaling up at speed. The following subsections discuss different implementational aspects of the system, including planning, financing and monitoring.

5.6.1 Implementing partners

The state has centred local and active partnerships between ULBs, Mission Shakti groups, TGs, waste pickers and urban residents in its implementation strategy, consciously steering away from PPPs and industrialised sources of the SWM-S. The paradigm shift has entailed greater governmental ownership over the design and implementation of the SWM-S.

The primary partners responsible for the functioning of the city-level SWM-S are:

(a) *Mission Shakti groups and TGs*: The state of Odisha launched Mission Shakti as a programme to unlock women's empowerment in 2001. Mission Shakti SHGs are the cornerstone of the programme, which also promotes the gainful employment of the WSHGs under the state's developmental activities and projects. By 2019, the state had enabled the organisation of 7 million women in 600 000 Mission Shakti WSHGs. A dedicated department with the state government, the Department of Mission Shakti, oversees the management of the overall programme.
(b) *Swachha Sathis and Swachha Supervisors*: The Swachha Sathis, members of the local Mission Shakti groups, have been roped in as community mobilisers dedicated to effect behaviour change in the households, of handing over segregated waste to the waste collectors – the foundation stone of decentralised SWM. Swachha Sathis under the supervision of Swachha Supervisors (one Swachha Supervisor engaged for four Swachha

Sathis) generate awareness on the best practices in SWM among urban residents, including collection of user fees under SWM.

(c) *Swachha Karmis*: Swachha Karmis, who are also WSHG/TG members, are the key performers responsible for collecting segregated municipal waste from each household through BoVs following the prescribed route, transporting it to the wealth centre and processing the wet waste in the MCC and dry waste in the MRF. A team of about 10 Swachha Karmis process the segregated waste at 5 TPD MCCs and 5 TPD MRFs.

(d) *Waste pickers*: Waste pickers are wandering informal waste collectors who go door-to-door buying recyclable waste – such as cardboard, glass, metal and so on – directly from households or pick up recyclables from the garbage dump yards and sell them to the traders to make a living. Under Odisha's SWM model, waste pickers have been identified, enumerated using a dedicated app, issued ID cards and integrated into the SWM value chain such that they engage with it in a dignified way to earn their livelihood. They are now provided with uniforms and safety gear, ensured dignity at the workplace, have access to various social welfare benefits under different government schemes and earn decent wages.

(e) *Cement factories*: On the advice of Odisha's State Pollution Control Board, four cement factories in the state that are able to co-process non-recyclable dry waste have been tagged to a specific nodal ULB, which in turn is tagged to other ULBs in its region. The state chose such a hub-and-spoke model to minimise the need for day-to-day coordination and communication between individual ULBs and the cement factories for the successful management of non-recyclable waste. In a highly streamlined operation, the nodal ULB coordinates with the tagged ULBs for the cumulative collection of non-recyclable waste and is responsible for its final conveyance to the designated cement factory in a coordinated manner.

(f) *'Ama Sahara' mobile app*: 'Ama Sahara' mobile app has been developed to facilitate:
  - data entry on a day-to-day basis on vital aspects in the wealth centres by the facility managers;
  - lodging of feedback or complaints by citizens pertaining to SWM;
  - payment of user fee through the payment gateway in the app.

The app provides interfaces for actors across the SWM value chain, enables the local and state governments to monitor and manage system-wide performance and helps in decision-making for effective management of the wealth centres. It fulfils the dual goals of operational management and monitoring. With respect to the former, it lets the facility in-charges log in information about the day's haulage of solid waste, the total quantity of waste processed and its disposal through various means depending on the type, and the revenue generated from the sales of compost and recyclable non-biodegradable waste. Secondly, it allows citizens to make payments for SWM services online and request garbage pick-up or orientation on segregation.

It also provides them the ability to lodge their complaints and any feedback relating to the services in an easy, transparent and quick manner for the ULB's action and overall improvement of the SWM-S. To aid the monitoring of operations, the app lets the state- and ULB-level administrators evaluate key performance metrics through a dashboard based on the incoming data (from citizens and facility in-charges).

5.6.2 Operations under a decentralised SWM-S

The organisation of the citywide SWM-S in urban Odisha follows a hierarchy. A city can have multiple clusters depending on its population, and each individual cluster follows a set pattern.

(a) *Sectors*: Sectors are the bottom unit of the set-up and represent the service area of a single collection vehicle. Each sector is served by two Swachha Karmis performing daily rounds of solid waste collection and one Swachha Sathi who works towards the awareness creation and compliance of communities and residents towards SWM-S.
(b) *Circle*: Four sectors together make a circle overseen by a Swachha Supervisor.
(c) *Cluster*: Three individual circles combine to form a cluster (Figure 5.3).
(d) *Wealth centre*: Each cluster has one wealth centre. Accordingly, the cluster-level wealth centre for solid waste collection, processing and management has a total of:
- three circles
- 12 sectors
- three supervisors
- 12 Swachha Sathis

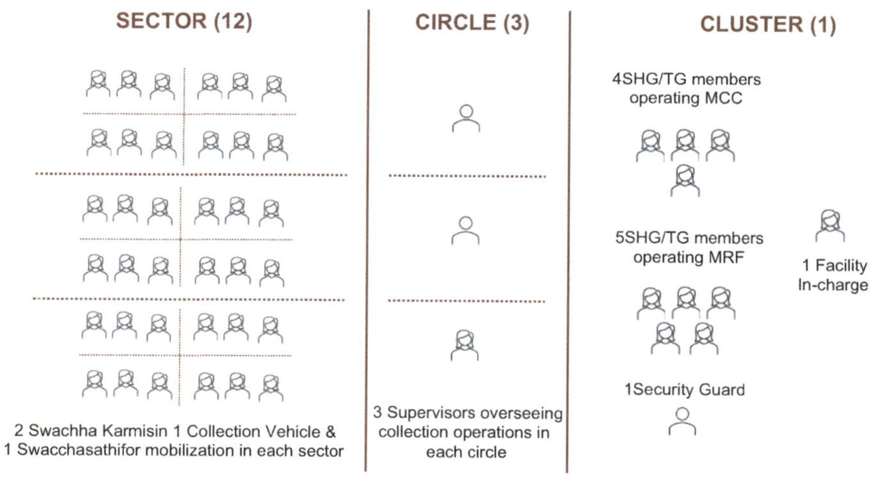

Figure 5.3 Organisational hierarchy of the SWM-S at the cluster level. *Source*: author's analysis.

- 24 Swachha Karmis
- nine members of WSHG/TG, together with a facility in-charge and a security guard.

5.6.3 Expenditures and revenues

Financial sustainability has often been the Achilles heel of urban services in India. Planning processes for infrastructure, while focused on matching the estimated capital expenditure to the available budget, can side-line the subsequent O&M costs the system is bound to incur. The failure to account for these can lead to a disruption of operations and, ultimately, a failure of the system in the medium and long terms.

Odisha has mobilised funds from within its budget to support the rapid scaling up of infrastructure and assets across all the ULBs. To tackle the challenges to sustainability of SWM-S, the state emphasised the evaluation of both the costs to implement and operate the system, and the different revenue streams that could contribute to cost recovery.

(a) *Expenditures*: There are three main expenditure heads.
  - *Infrastructure:* The facilities created for SWM, that is the wealth centres consisting of MCCs and MRFs, need CAPEX for building infrastructure.
  - *Procurement of equipment:* Cost of procurement of equipment – for example, shredding machine, sieving unit, weigh bridge, fly catchers, incinerator, baling unit and so on, BoVs/LCVs – is also included in CAPEX. As per the state's estimates, a wealth centre with a capacity of 5 TPD, capable of serving a population of 33 000, has a capital cost of 11 million (USD 148 483). Multiple wealth centres have been set up based on the population, requirement and availability of space in a ULB.
  - *O&M:* Being a low-cost technology, incentives for the SWM-S personnel constitute 85% of the system's total operating costs. The remaining expenditure goes towards energy bills, repairs, maintenance, tools and ingredients for effective microorganism solutions. For a population of 33 000, the operation of the SWM-S entails an annual O&M expenditure of Rs 7.21 million (USD 192 674).

(b) *Revenues*: The vital components are:
  - Collection of user fee under SWM. Households pay a minimum of Rs 1 per day, or Rs 30 per month (USD 0.4 per month) to avail the SWM services.
  - Sale proceeds from 'Mo Khata'. The organic compost from the MCC, or Mo Khata, retails at Rs 20 per kg and, with a cost of Rs 5 per kg for bagging and labelling, brings in a profit of Rs 15 per kg (USD 0.2 per kg) sold.
  - Sale proceeds from dry waste. The wealth centres sell the recyclables to *kabbadiwalas* at an average rate of Rs 6 per kg (USD 0.08 per kg) for a significant share of the total revenue.

The revenue resulting from the sales of compost and recyclables and collection of user fees creates a steady inflow to support operations.

5.6.4 Monitoring SWM through the Ama Sahara app

Monitoring and evaluation are indispensable cornerstones of any successful intervention. The global mainstreaming of GovTech[3] in the last few years has made digital tools common and easily accessible solutions for the integrated management of day-to-day operations and their monitoring at different hierarchical and temporal scales. Odisha has adopted the digital route to monitoring SWM-S with launching the 'Ama Sahara' mobile app and 'Swachha Sahar Odisha' web app on 31 August 2020, the state's Local Self Government Day.

The scope of Ama Sahara is being expanded to make it a complete monitoring tool for SWM-S across ULBs, district and the state level.

5.6.5 Urban–rural convergence for state-wide SWM

Contrary to popular imagination, the administratively 'rural' areas of the state and the country at large comprise a diverse set of settlements. Although the smaller and predominantly agrarian villages produce minimal non-biodegradable waste and recycle biodegradable waste as compost, the larger and denser villages (including census towns) are increasingly emerging as hotspots of non-biodegradable waste. In the absence of formal rural SWM-S, these areas have had to contend with littering in public spaces, dumping of waste in the open environment, and its indiscriminate burning.

The central government is attempting to resolve rural India's challenges in SWM through the second phase of the Swachh Bharat Mission-Gramin. The programme provides financing worth Rs 1.6 million (USD 21 562) per block to district administrations for the development of a rural MRF or plastic waste management (PWM) unit. Foregoing greenfield development by building standalone rural MRFs, Odisha has demonstrated a resource-efficient alternative by leveraging urban MRFs to cater to the rural population's PWM needs. Department of Housing and Urban Development (HUDD) and its rural counterpart, the Panchayati Raj and Drinking Water Department, signed a Letter of Understanding with UNICEF and SCI-FI on 21 September 2021 for taking up this urban–rural convergence. Consequently, all 30 of the district administrations in the state have tagged gram panchayats (GPs) under their purview to the nearest or most suitable ULB.

The rural local bodies and ULBs are responsible for the successful implementation of the rural–urban regional model for managing dry waste. The GP administration is responsible for the collection of dry waste from rural households and its transportation to the MRF(s) of the ULB to which it is tagged. Where the daily volumes of the collected waste are low, the state

---

[3] GovTech is a whole of government approach to public sector modernisation. GovTech emphasises three aspects of public sector modernisation: citizen-centric public services that are universally accessible, a whole-of-government approach to digital government transformation and simple, efficient and transparent government systems.

has encouraged GPs to construct storage sheds for temporarily containing the cumulative waste before its transportation. Once the waste reaches the MRF(s), the ULB is responsible for ensuring it is safely processed and disposed along with the rest of the dry waste from urban areas. Although the GPs incur the expenditure for the collection of waste and its transportation to the MRFs, the ULBs process the waste at their own cost and, accordingly, are not required to transfer any revenue they earn to the GPs.

5.7 PROGRESS ON THE GROUND

As on July 2021, nearly a year after the launch of the SWM campaign, residents in 100% of wards across all ULBs in the state have access to door-to-door waste collection services, and residents in about 90% of the wards are practicing source segregation. More than the state's prowess to scale up SWM infrastructure, the progress speaks on the tremendous success in changing urban dwellers' behaviour towards a public good. In terms of infrastructural scale-up, starting from zero waste processing in early 2020, the state now has approximately 759 and 1564 TPD of biodegradable and non-biodegradable waste processing capacities, respectively.

In the process, ULBs have cumulatively partnered with 1785 Swachha Karmis, 2120 Swachha Sathis and 485 Swachha Supervisors through Mission Shakti groups (both WSHGs and TGs). Additionally, ULBs have enumerated 2170 waste pickers and integrated them into the formal system for SWM. So far, the system has generated a total revenue of Rs 62.2 million through user fees under SWM, and the sale of Mo Khata and of non-biodegradable recyclables. Because of the decentralised nature of the system, the profits and benefits generated by the SWM-S are directly distributed among the local communities that are responsible for its management. ULBs have issued standing instructions to the banks for remittance of the monthly incentives by the 7th of each month (direct benefit transfer) to the accounts of the community partners, unless 'stop payment' instructions are specifically issued in anyone's case. Thus, the 'wealth from waste' generated in the wealth centres flows to the community partners – a distinct manifestation of a 'circular' economy.

5.8 WELL BEGUN IS HALF DONE

Odisha's SWM-S has successfully eliminated the problem of 'ghost garbage' and brought in attendant cost savings. Before its implementation, ULBs in Odisha spent about Rs 390 million every month just for the handling of garbage, meaning it was just moved from one location to another. The SWM-S has not only staved off this expenditure, but also created a cash flow that is likely to be instrumental in keeping it going.

Odisha's SWM system has also successfully integrated local communities and community organisations into a network of practicable public activity. The integration of not just WSHGs and TGs but also of waste pickers provides a holistic framework that could be replicated without high costs or lots of training. The space available for modifications based on local inputs and the structural provisions for continuous feedback also clearly demonstrate that decentralisation can truly aid implementation.

All this, combined with its deliberately low-tech approach, imparts Odisha's SWM-S an inherent modularity that could make it a starter kit for implementation across a whole range of contexts, including in rural areas. The key transformation is that 'community' replaces 'contractors', with the driving force for the transformation being the community partners. This unique, dynamic model of decentralised SWM has led to real empowerment for women, transgenders are finding a new dignified identity and waste pickers have been integrated into the SWM value chain with safety and dignity ensured.

In conclusion, the Odisha's SWM model is a dynamic framework poised for continual enhancement. Looking ahead, a key focus will be on fortifying the existing system. Transforming community partners from service providers into micro-enterprises, in collaboration with private enterprises, will harness private sector expertise for more efficient recycling. Additionally, expanding the SWM-S to embrace the 3R concept – reduce, reuse, recycle – by establishing 3R centres in all ULBs will demonstrate these principles to the public. These initiatives are essential steps towards achieving the vision of 'zero-waste cities', grounded in the principles of a circular economy, technological innovation and sustainable, community-led systems.

Before SWM

Source: HUDD, GoO.

MCC MRF Hinjlicut

Source: HUDD, GoO.

Solid waste compactor

Source: HUDD, GoO.

Recycling of solid waste

Source: HUDD, GoO.

doi: 10.2166/9781789064926_0075

Chapter 6

Tap water supply to every home and beyond

ABSTRACT

This chapter presents an account of the various initiatives undertaken by the Housing and Urban Development Department, Government of Odisha to achieve the goal of ensuring universal access to reliable and good quality tap water supply for all urban households in the state.

Keywords: Universal access, drink from tap, metred connections, community engagement, public health, Jal Sathi, water-borne diseases, revenue recovery, BASUDHA, AMRUT, finance commission.

6.1 W(H)ETTING AN APPETITE

Water is critical for human life. Even in the ancient world, no shelter, irrespective of the safety it offered, was feasible without a source of water nearby. In the centuries since, access to water has led to a visible imposition of hierarchies in the way cities have been laid out. In the modern world, running water supply inside a house has become a fundamental aspect of urban life. The advancement of a society has also at times been linked to the quality of water that flows through those pipes.

Safe and readily available water is important for public health, whether it is used for drinking, domestic use, food production or recreational purposes. Improved water supply and sanitation, and better management of water resources, can boost countries' economic growth and can contribute greatly to poverty reduction. When water comes from improved and relatively accessible sources, people spend less time and effort physically collecting it, meaning they can be productive in other ways. This can also result in greater personal safety and reduce musculoskeletal disorders by eliminating the need to make long or risky journeys to collect and carry water. Better water sources also mean less expenditure on health, as people are less likely to fall ill and incur medical costs; they are also better able to remain economically productive. With children particularly at risk from water-related diseases, access to improved sources of water means better health, and therefore better school attendance, with positive longer-term consequences for their lives.

Starting from a low base, India has made huge strides in ensuring that its households have access to piped water. However, efforts are continuing, because access is still not universal. Census figures show that 68.7% of urban households in India had access to piped water in 2001, with the number rising to 71% in 2011. NITI Aayog's composite water management index, 2019[1] showed that 93% of India's urban population had access to 'basic water supply'.

However, the water that is indeed being supplied is often not up to the mark. A study in 2019 by the Bureau of Indian Standards (BIS)[2] showed that 173 out of 210 (82%) water samples taken from the national capital and 20 other state capitals did not meet the quality standards. The lack of adequate and good quality water in turn leads to deficiencies in sanitation and health. Stressed water sources can often lead to outbreaks of diseases such as diarrhoea, jaundice and cholera. India's breakneck urban expansion over the past few decades has seen cities grow with severe water crises. Of the 20 largest cities in the world that are water stressed, five are Indian, with Delhi ranking the second. Also, a total of 756 urban local bodies (ULBs) have been declared water stressed by the Union Ministry of Jal Shakti.

Apart from the loss of water bodies to urban expansion and the growing impact of climate change, the management of the water supply value chain has proven incapable of meeting the demands of India's urbanisation and its sustainability. Additionally, water utilities across the nation has been hobbled by inefficient operations arising from various factors leading to poor service delivery and very low level of public satisfaction.

6.2 ODISHA: DROP IN AN OCEAN

During 2014–2015 only 40% of the total households in Odisha had access to piped water supply. The average per capita supply of water at that time was reported to be 60 litres per capita per day (LPCD) against the national benchmark level of 135 LPCD. However, this was an incorrect depiction as the estimated figure of 60 LPCD supply was based on the entire city's population instead of the population in the piped network area. This implied that more water was being provided to the 40% area covered by the piped network whereas the remaining

---

[1] The National Institution for Transforming India (NITI) Aayog has developed the composite water management index (CWMI) to enable effective water management in Indian states. The CWMI is the first comprehensive collection of country-wide water data in India based on in-depth structured questionnaires followed by focus group discussions to generate qualitative information.

[2] https://bis.gov.in/wp-content/uploads/2019/11/Drinking_Water_20112019.pdf.

uncovered 60% area was fed only by handpumps, tube wells or polyvinylchloride tanks filled through water tankers. In such scenarios, people have no other option than to fend for themselves. Persistent scarcity and irregular availability of water in non-networked areas help private water traders to flourish.

Quantity of water supply was higher in larger ULBs than the smaller ULBs and the ULBs which were far away from the state capital. There existed a huge disparity in quantum of per capita supply between the cities and areas within a city. For example, Bhubaneswar was getting an average of 350 LPCD supply whereas the water supply in many other cities was even less than 40 LPCD. Similarly, even within a city such as Bhubaneswar some areas were getting more than 800 LPCD, whereas in other areas it was as low as 40 LPCD. Also, total supply at the state level fell short of demand by 157 million litres per day (MLD).

Odisha has also suffered from outbreaks of diseases such as jaundice, diarrhoea and hepatitis in Puri, Bhubaneswar, Cuttack, Sambalpur and Bolangir in recent years, due to deficiencies in the quality of water supply. During 2015, the state had declared the districts Sambalpur, Bolangir, Cuttack, Khurda and Jajpur jaundice affected. Sambalpur had reported 2945 and Cuttack had reported 116 cases; the total number of cases in the state was 3966. The official death toll was 36 (Figures 6.1–6.3).

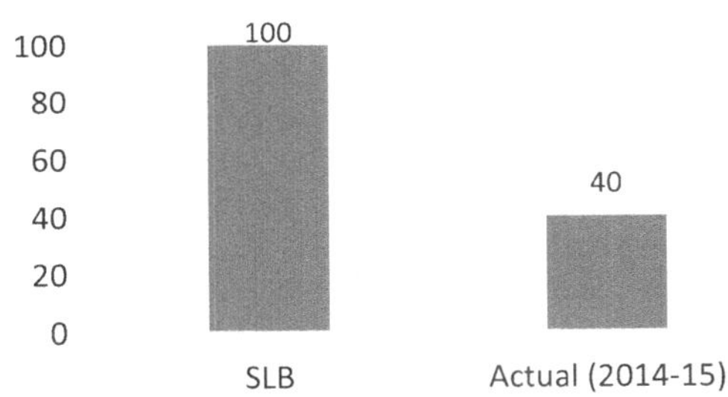

Figure 6.1 Coverage with piped water supply systems (% households), 2014–2015 vs service level benchmark. *Source*: Service Level Improvement Plan (SLIP) under AMRUT 2017–2018.

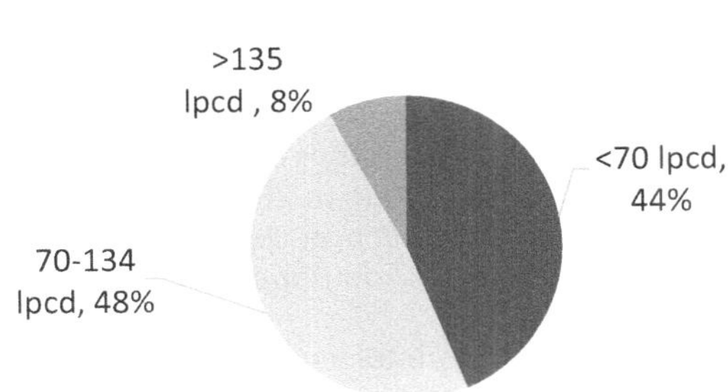

Figure 6.2 Distribution of Odisha's ULBs according to quantity of water being supplied in terms of LPCD (2014–2015). *Source*: Service Level Improvement Plan (SLIP) under AMRUT 2017–2018.

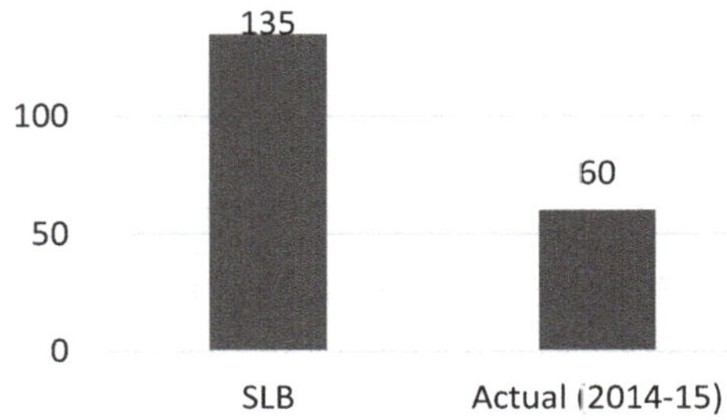

Figure 6.3 Availability of water (in terms of LPCD), 2014–2015.

6.3 UNIVERSAL ACCESS TO PIPED WATER SUPPLY PROGRAMME

In this background, the Housing and Urban Development Department (HUDD), Government of Odisha, launched a programme Buxi Jagabandhu Assured Drinking Water to All Habitations (BASUDHA) – universal access to piped water supply programme aimed at ensuring universal access to clean, equitable, safe and affordable piped water supply in all ULBs. The programme was conceptualised based on the premise that universal coverage is a prerequisite for optimising the public health benefits of improved water supply. The state government believed that to achieve the goal of universal coverage, the first step was to prioritise inclusion of the currently excluded settlements and communities (slums and other informal settlements) irrespective of their land tenure and legal status. The programme was thus aimed at covering all wards/areas (including slums and informal settlements) across the 115 ULBs in the state.

To achieve the goal of universal access to piped water supply for all urban areas, well-defined time-bound targets for service delivery improvements were set and a phased implementation approach was adopted.

On 25 June 2015, the Government of India launched the Atal Mission for Rejuvenation and Urban Transformation (AMRUT). The national hierarchy of priorities stipulated by the AMRUT mission is non-negotiable, which means that funds provided under AMRUT must necessarily be spent in the selected cities for the selected schemes in the predefined order of priority. Until all homes in all AMRUT cities are provided with piped water supply connections, for example, the scheme funds cannot be utilised for any other purpose in any AMRUT city.

AMRUT was launched in 1 lakh-plus cities in India, including nine in Odisha. The initiative was kickstarted in the state in 2016 with the inauguration of 300 water supply projects (100 projects each in April, June and October). In the same year, work was initiated for an additional 150 water supply projects (50 projects in July and 100 in October).

AMRUT has given a mandate to implement the piped water supply scheme in all the nine AMRUT cities in Odisha on a saturation basis, prohibiting diversion of funds for any other purpose in any other city until saturation in all cities. This novel and unique aspect of the scheme fascinated me the most. Top priority given to piped water supply by AMRUT helped to drive the universal tap water supply agenda of the state.

With the launch of AMRUT in 2015, the implementation of Odisha's water policy of 2013 – which envisaged providing adequate quantity of water to all urban residents by 2027 – was stalled. Learning from the insights of the universalisation approach and the topmost position for water in AMRUT's priority list, the state envisioned a new programme called 'Buxi Jagabandhu Assured Drinking Water to All Habitations (BASUDHA) – universal access to piped water supply programme' on the lines of AMRUT to cater the requirements of the entire state. BASUDHA, launched in 2017, revised the targets of the 2013 water policy advanced by 4–8 years. With respect to water availability the targets were brought forward by 4 years (from 2023 to 2020), and those pertaining to physical and household coverage advanced by 8 years (from 2027 to 2019) (Table 6.1).

6.3.1 Launch of BASUDHA

The year 2017 was a landmark for the BASUDHA initiative as 384 water supply projects were inaugurated (in April and October) and on 20 October the BASUDHA initiative was launched. BASUDHA, under state funding, aimed to ensure 100% piped water coverage in all wards (including slums) in the entirety of urban Odisha covering all ULBs and to increase the quantity of water supply to a minimum prescribed limit of 70 LPCD within 2 years (by October 2019). In the year of its launch, work on 555 water supply projects was initiated; 500 water supply projects were commissioned in November 2018 by HUDD, Government of Odisha (Figure 6.4).

Table 6.1 Phase-wise targets for water quality, quantity/availability and coverage under AMRUT (2015–2027).

Parameter	Unit of Measurement	Status (2014–2015)	Target		
			1st Phase (2015–2019)	2nd Phase (2019–2023)	3rd Phase (2023–2027)
Water quality	Test passed	80%	100%		
Water availability (70 LPCD)	No. of cities	55%	100%		
Water availability (135 LPCD)	No. of cities	10%	70%	100%	
Physical coverage	Area of the city	60%	75%	94%	100%
Household coverage	House connections	25%	60%	80%	100%
Metering	House connections	0%	70%	100%	
Non-revenue water	Water produced	50%	30%	19%	15%
Energy efficiency	No. of cities	5%	60%	100%	
Cost recovery	Revenue to O&M cost	35%	50%	80%	100%
Supply duration	Hours per day	2.4	4	8	12

Source: SAAP 2017–2018.

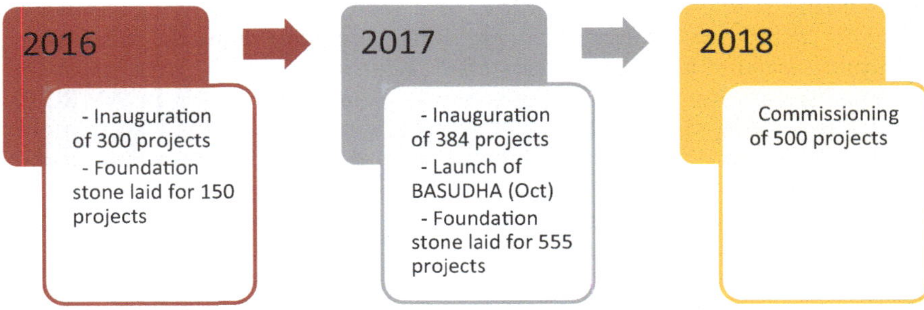

Figure 6.4 Timeline for initiation, and commissioning of water supply projects under the universal access to piped water supply programme. *Source*: authors' analysis.

Around 1200 projects were commissioned between 2017 and 2023 with an investment of approximately Rs 60 000 million. An investment of Rs 40 000 million in infrastructure upgradation was made between 2017 and 2021, amounting to a five-fold increase in the annual investment in the urban water supply sector – from Rs 2000 million to Rs 10 000 million. This has subsequently increased to Rs 20 000 million in 2022–2023 and Rs 30 000 million in 2023–2024 with implementation of drink-from-tap (DFT) projects. The water supply distribution network more than doubled – from 5383 to 11 325 km – which resulted in a surge in network coverage areas from 40% in 2015 to more than 90% in 2020. Forty-nine towns that were water deficient in supply had turned water sufficient, supplying 70–135 LPCD to households. Moreover, house connections increased from 0.3 million to 1.04 million – a jump of about 250%. Additionally, nine water testing laboratories were established across the state to monitor the quality of water supply.

In October 2020, history was created in Bhubaneswar when it became the first city in the country to achieve the milestone of 100% coverage of houses with piped water connections. With 105 out of 115 cities achieving 100% house connections by March 2024, the government is on track to achieve this target across all the cities of the state.

6.4 KEY FEATURES OF THE UNIVERSAL ACCESS TO PIPED WATER SUPPLY PROGRAMME

6.4.1 Small- and medium-sized projects

Learning from AMRUT's implementation, small- and medium-sized projects – instead of large contracts of higher value – were formulated for the following reasons:

- Local contractors are rendered ineligible for participating in tenders of high-value projects, and the contracts are garnered by bigger players, mostly from outside the state. Besides limiting competition, the high-value tenders are mostly taken at high premium defeating the very purpose of tendering.

- When disputes arise in the execution of larger projects, the litigation process stalls the execution of the entire project until disposal of the litigation. When the full project is sliced into smaller independent projects, disputes arising in one do not affect the execution of the rest.

Keeping these facts in mind, small- and medium-sized projects were tendered and implemented locally at the level of Executive Engineer, mostly with the local contractors, with an implementation period of 6–12 months.

6.4.2 Contractors from different sectors

To decide the size of the contract and number of projects that can be implemented in every division, availability of division-wise local contractors was assessed. The assessment led to the realisation that many divisions had very few contractors, and with limited financial capacities. As this situation might have led to tender cartelisation, and implementing many projects with these few local contractors was impractical, the government decided to look for additional contractors engaged in other sectors and departments, such as roads, water resources and so on, across the state.

A consultative workshop was organised with the contractors' associations and their members to seek suggestions on bid criteria that would enable their participation in the tenders. Thereafter, the standard bidding document was suitably amended to facilitate the participation of contractors from other sectors. This resulted in onboarding of many new contractors for water supply projects across the state, successfully scuttling cartelisation, and facilitating increased competition and faster implementation.

To achieve piped water supply connection to every home in 3 years' time, the government opted for a massive, piped network infrastructure for distribution. As regards supply, wherever river sources were expected to take longer time for execution, as interim solutions, the state went in for production wells and rationing of the existing supply quantity to ensure equitable distribution across a city with the increased pipe network. Large numbers of contractors were engaged to ensure that the piped water supply network projects were completed in a period of 6–9 months.

6.4.3 Better detailed project reports, prepared in-house

The next issue to be resolved was the in-house capacity constraints at the departmental engineer's level. Traditionally, detailed project reports (DPRs) are prepared for infrastructure projects by engaging consulting agencies due to lack of in-house capacities. This practice had several disadvantages, such as limited availability of consulting agencies, longer time taken for preparation of DPRs, consulting agencies resorting to desk-based DPRs without necessary field verification, preparation of unrealistic estimates and so on. In addition, due to lack of in-house capacities, the DPRs prepared by external agencies were getting approved without due scrutiny by the Public Health Engineering Organisation

(PHEO).[3] Adopting such DPRs often resulted in higher tender premium – as high as 80–90% in some cases – forcing the department to refer the matter to Finance Department (FD) for concurrence causing considerable delays. This was the scenario when the PHEO was executing fewer projects at the divisional level.

Upon the launch of BASUDHA, a new strategy to address these issues was adopted. Learning from Tamil Nadu, software is now used for the preparation of DPRs, which has made the process easy, accurate and faster. A team of around 30 engineers were deputed to Tamil Nadu Water Supply and Drainage Board to undergo training. The team that was trained came back and trained other engineers in the PHEO divisions.

Subsequently, the engineers started preparing DPRs of much better quality and speed than the consulting agencies. This capacity building effort led to in-house preparation of all DPRs without the need for engaging consulting agencies for DPR preparations, saving time and money. The familiarity of the state's engineers with the field conditions enhanced the accuracy of the DPRs resulting in more realistic estimates. This resulted in more competitive bids and reasonable tender premiums – in the range of ±3%. These measures have not only saved time and money but also circumvented the need to take FD's concurrence for every tender.

6.4.4 Procurement of materials by the government

Another critical strategy that has led to faster implementation of the programme was reversing the embargo on departmental procurement of pipes imposed by the government a decade ago. This ban had been imposed by the government to curtail scams involving needless procurements of pipes during financial year-ends merely for the purpose of utilising budget allocations and to avoid surrender of funds; these pipes would remain stacked in piles in stores without creation of a water supply facility. In a distribution network project, the material component is about 70%, with pipes and fittings as major constituents; the labour component is about 30%. The contractor is compelled to invest 70% of the cost in the procurement of pipes, which must be done in advance, and it takes longer for him to get paid for materials after completion of the work.

It was realised that if procurement of materials is taken away from the scope of the contract, then the cost of the contract is reduced substantially, allowing more contractors to participate in the bidding process. This also facilitates easy mobilisation of workforce and resources for the contractor, leading to faster implementation. Reduction of contract value also enables the existing local contractors to bid for multiple projects leading to more competition and lower tender premiums.

6.4.5 Streamlining processes

Having knocked off the time delays that could happen in procurement, the government started identifying the process delays and ways to fix them for barrier-free speedy implementation of the projects.

[3] The main aim and objective of the PHEO is construction, commission, inspection and monitoring of water supply systems of all urban local bodies of Odisha.

Separate administrative approval for each project was one major source of delay. Each project required the DPR to obtain administrative and technical approval before it is tendered for selecting an executant. After approval of the action plan with the list of projects at the governmental level, DPRs of individual projects were scrutinised for administrative approval – a bureaucratic decision-making process involving commitment of funds and approval to execute the project. Any water supply project of value up to Rs 150 million was approved at the level of the minister in-charge of the department. Projects of more than Rs 15 million value required FD concurrence before they were approved by the minister. This process of obtaining administrative approval involved many layers of the government, both in HUDD and FD, and usually took several months. Administrative approval of, say, 555 projects in this conventional way would have involved a lengthy bureaucratic process of dealing individually with 555 files, resulting in abnormal delays even before commencement of the tendering process.

Odisha therefore decided to depart from the normal process of floating 555 individual files for receiving administrative approval for the 555 projects and went for a composite administrative approval at the minister level with FD concurrence through a single file clubbing all the 555 projects together. This file process reengineering saved the state substantially from unnecessary file work and resultant delays.

6.5 DFT MISSION

Although the state was successfully implementing the universal access to piped water supply programme, in the monsoon of 2019, sporadic incidences of diarrhoeal and jaundice cases were reported. Although there were no deaths linked to them, these incidences triggered the question as to the actual quality of water supplied. Hence, the HUDD shifted its focus to ensuring good quality water supply.

One of the prerequisites for preventing contamination in the transmission and distribution network is supply of continuous pressurised water instead of the prevailing intermittent supply. Supply of continuous pressurised water on a 24×7 basis has the potential to result in huge household savings, as the individual households will no longer require water storage tanks and water filtration (reverse osmosis) systems. Inadequate and inferior quality of water supply forces people in the low-income group (LIG) and middle-income group (MIG) segments to spend a sizeable portion of their monthly household earnings in buying drinking water. For most of the urban poor, including slum dwellers, buying good quality water is not affordable and they resign themselves to the inadequate and inferior quality of water supplied. After weighing the merits and advantages that would ensue from 24×7 water supply of directly drinkable quality and the huge public benefits – including improved public health outcomes – this would usher in, the government decided to set its sights on enhancing the quality of the water supplied. Although supply of improved and standardised quality water would benefit the entire urban population, the impact was likely to be greater on the lives of slum dwellers, urban poor and

LIG and MIG segments which constitute most of the urban population. Thus, the DFT concept emerged as the next mission in Odisha's water journey. The 'Drink-from-Tap (DFT) Mission' was launched in 2019 as a pilot initiative.

The DFT mission is aimed at ensuring safe and continuous (24 × 7) water supply through 100% metred connections. By providing continuous water supply, the scheme aims to eliminate any possibility of contamination of water either during transmission (due to intermittent supply) or during storage at the household level (in overhead tanks/underground sumps). The DFT mission is aimed at ensuring that drinking water complying with IS 10500 standards is provided to each home and consumers can drink water straight from the tap without the need to filter or treat water prior to consumption.

The implementation was entrusted to a newly created corporate entity, the Water Corporation of Odisha (WATCO). A core team was formed in WATCO with engineers of the state government who had exhibited extraordinary commitment and passion and had delivered exceptional performances in the past.

During brainstorming sessions to discuss the way forward, two more fundamental issues came up: tackling high water losses and ensuring equity and consistency in supply of DFT quality water at every home at every tap. Although the demand vis-à-vis production of water was at par, as per a water audit study conducted in 2015, it was found that unaccounted-for water or non-revenue water (NRW) in Odisha was a whopping 54% of the produced water against the national level benchmark of 20%. NRW is the water that has been produced and is 'lost' before it reaches the customer. Losses can be real losses through leaks (also referred to as physical losses) or apparent losses (through thefts or metering inaccuracies). Considering the urgency to transit to 24 × 7 water supply on a sustainable basis, the issue of NRW had to be tackled aggressively.

The other issue to be addressed was the installation of own underground/ overhead storage tanks with pumps by individual households, through which water was illegally drawn from the piped water network during supply hours. This created inequity as households in the downstream seldom got adequate water supply. It is common practice for people to store more water than what is required and waste the excess stored water when fresh supplies are received. The more they store, the more they waste. Furthermore, there was chance of contamination of water in the household storage tanks, in the absence of regular cleaning of the tanks by the consumers. It was evident that by providing 24 × 7 water supply in a reliable manner, equity in distribution and prevention of wastage could both be achieved effectively.

As there was lack of prior experience in 24 × 7 water supply systems, the government decided not to go all out but instead experiment with the mission in a smaller area; this could then be scaled up to cover cities, depending on the learnings. The pilot initiative was implemented in eight water supply zones in Bhubaneshwar and four in Puri. The selection criteria adopted for shortlisting water supply zones for the pilot initiative were as follows: the area must have a mixed population with at least 20% of the population/households belonging to the economically weaker section; it should have around 1500–2000 consumers and it should have a relatively new piped water supply network. The selected water supply zones were clearly demarcated to ensure there was no disruption of water supply services in adjacent zones.

The choice of the areas to be covered by the pilot initiative was strategic: these were high-visibility areas and success would ensure support from other ULBs and citizens elsewhere. Ishaneswar Basti in Bhubaneswar served as a living laboratory for the implementation of DFT, helping to develop the basic yet full-fledged implementation framework piloted in a few zones of Bhubaneswar and Puri cities for fine-tuning prior to upscaling. A systematic step-by-step approach, with a steady learning and doing curve – moving from the laboratory to the ground – was ensured. The pilot zones in Bhubaneswar included Salia Sahi, the largest slum in Odisha with a population of 57 500 people. Developing the water supply infrastructure for Salia Sahi included laying of distribution mains in uncovered areas, covering narrow lanes and by-lanes, setting right the leaky ferrule connections, setting up systems in kutcha dwellings, laying pipelines under concrete roads and pavements, construction of elevated storage reservoirs (ESRs), ensuring 100% connection to all households, metering of the connections, installing chlorine dosers and analysers using programmable logic controllers for real-time monitoring and so on. Even as distribution pipelines with their accessories were being laid, all infrastructure assets in the area, existing and new, were mapped onto a geographical information system (GIS) platform for use during execution as well as operation and maintenance (O&M).

The pilot initiative in Puri was soon scaled up to cover the entire city. In July 2021, Puri became the first city in India to achieve DFT quality water supply with 100% metred connections, thereby joining cities such as London, New York and Singapore. The programme has benefitted around 0.25 million residents of Puri including 66 000 urban poor as well as 20 million tourists who visit this heritage city every year. The project has the distinction of being commissioned within 9 months of its initiation.

To reduce plastic waste in the city, drinking water fountains have been installed at key public places. It is estimated that these drinking water fountains will help reduce use of plastic mineral water bottles eliminating 400 MT of plastic waste every year.

Furthermore, in order to ensure monitoring of water quality at the local/community level the mission has collaborated with Jal Sathis,[4] the women's self-help groups (WSHGs) created under Mission Shakti. The Jal Sathis were provided with water-testing kits and trained to carry out household-level water quality indicator tests.

After successful implementation of the pilots and commencement of city-wide scaling up in Puri, planning for the upscaling of DFT in other cities of the state was carried out. As the Odisha Mineral Bearing Areas Development Corporation (OMBADC) had huge funds available for investments in bigger projects, it was proposed to take up universal coverage of piped water supply

[4] 'Jal Sathi' initiative to engage the women from self-help groups in water distribution and consumer management with the aim of women empowerment.

in 14 cities that fall under the OMBADC area with OMBADC funds. On 27 December 2019, OMBADC's Board issued in-principal approval for overhauling of water supply systems in the 14 cities to provide safe and clean water supply to every house, making a tentative allocation of Rs 12 000 million for the projects. A DPR was prepared for the provision of 24 × 7 supply in all the 14 OMBADC cities and tenders were floated. Providing 24 × 7 DFT in mineral-bearing underdeveloped districts is going to set a new benchmark for the mineral-bearing areas across the country. This is expected to benefit 1.6 million population living in the 14 cities including 0.4 million urban poor.

Ishaneswar Basti: implementation of DFT

On a Sunday morning, I, along with my team, went to Ishaneswar Basti (a small slum on a hillock with 471 households in Bhubaneswar) and climbed the reservoir to have a bird's-eye view of the slum. During our interactions with the residents, we witnessed with dismay the precarious and deplorable state of water storage and use at the household level. One of the factors we considered while selecting the sub-pilot area was to have habitation with limited households where water supply could be easily isolated. Thus, disruptions if any during this are limited to a smaller population which can be managed without causing inconvenience to the larger population.

After the gap assessment study and detailed asset and consumer survey of all 471 houses, a plan of action was prepared and the project in Ishaneswar Basti was set in motion with very little idea of the road and challenges ahead.

As this is a pioneer initiative in the country, we were on the look-out for global technical partnerships to bring in expert inputs and best practices. United Nation's International Children's Fund (UNICEF) was identified for technical assistance and a memorandum of understanding (MoU) was signed with them on 25 October 2019. Through the MoU, we received technical support from Mr Richard Francys and Mr Anand Jalakam. They shared valuable inputs on the learnings, chances and reasons of failures of 24 × 7 projects executed across the country. The UNICEF experts helped us with the review of the DPR, limited to design aspects of the 24 × 7 supply system. DFT called for in-house development of an indigenous system aided by our own hands-on learnings.

In November 2019, a selected team of six officials was sent to the Administrative Staff College of India (ASCI) for a 3-day training programme to understand the basics of 24 × 7. The team returned from the training fired up with the new learnings and insights and accelerated the implementation of the sub-pilot work in Ishaneswar Basti.

– G. Mathi Vathanan

6.6 KEY FEATURES OF DFT

This section outlines some of the main aspects and innovations of the 'DFT mission'.

6.6.1 Collecting infrastructure asset information

When the government embarked upon planning for DFT, it needed to find out the status of existing infrastructure, laid at different points of time and under different projects. Obtaining reliable data on such assets is always a difficult task as the water supply offices often lack organised record-keeping. Generally, record-keeping is poor in such engineering departments and the incumbents often get transferred or superannuated resulting in loss of institutional memory.

The issue of unavailable infrastructure asset information was managed through a variety of ways including finding the helpers, pump operators and linemen who had retired from the services as far back as a decade ago to get information about the pipe alignment and other asset details. This way more accurate information on the asset details was received than what was available in office records. In addition, the interactions with the local communities, especially the older people and retired linemen and pump operators were very helpful. With thorough field surveys, each asset was mapped over the GIS platform to identify the infrastructure gaps to be filled for DFT. Hence, interactions with the old grassroots-level field employees, community interactions and field surveys could bring in authenticity and completeness of the infrastructure asset data.

6.6.2 Consumer information availability

Unlike the banking sector where know-your-customer (KYC) is an established practice with the customer profile used for providing better services, in the utility sector including water supply this is not so. The details about house connections, consumption patterns, payment details and so on were not readily available when the Odisha government embarked on this scheme. There was a total absence of consumer connect, to the extent that the water supply officials did not know their consumers and the consumers hardly knew the water supply authorities.

The essence of DFT is the quality of water supplied to every home; therefore, knowing about the consumer is vital to establish consumer connect and improve the quality of service. Hence, it was felt that mapping the location and details of each consumer is essential for planning, designing, establishing and running a DFT system that meets the customer's satisfaction. Door-to-door consumer surveys and developing a GIS-based database of consumers with all required details helped collect actual field information with geographical locations and facilitated accurate planning and smooth implementation. In this way, the KYC of the consumers (existing and potential) for universal coverage and improved services was made available.

6.6.3 Overcoming resistance to house connection and metering

Taking a house connection was not acceptable to everyone. Two categories of people resisted house connections. One group was those who already had household borewells sunk by them at their cost. They operated their borewells as they wished and did not have to pay any recurring water charges. The second category consisted of those who had managed to acquire illegal connections and were enjoying free water. In addition, there were a few people in the water supply business who feared that their business would be affected if everyone took water supply connections. In addition, as it was made clear that as part of DFT every house connection would be metred, many feared increased monthly water charges and hence opposed metering.

The issue of resistance to house connections was tackled through creating public awareness and community engagement through Jal Sathis. In addition, policy measures such as providing the in-house connection by the water supply authorities, treating them as public work, introduction of an instalment system of payment for house connections and metering, making house connections free for urban poor, exemption from road-cutting charges and a hassle-free connection process contributed to achieving 100% house connection leaving no one behind.

However, a stiffer challenge was faced during the installation of water meters. People were used to paying fixed minimal charges with unlimited consumption. Therefore, the concept of metering was not well received. Consumers realised that with metering they must pay for their actual consumption. Contrary to the government's expectations, it was relatively easy to convince the slum dwellers on metering and payment based on actual consumption. The resistance was higher in the posh areas inhabited by the urban elites. It became evident that the consumption in urban poor settlements is not only low but also corresponded to the actual need, whereas in the higher-income areas, unmetered water supply was used for various purposes – including gardening, car washing and so on – with unaccounted wastages. Hence, those areas were more resistant to the installation of meters.

Intensive efforts were made to sensitise the households and to build local pressure through Jal Sathis and local community leaders. The fact that metering was made a non-negotiable factor and all households including the slum households are covered under metering without any exemptions made everybody accept the inevitable. In some areas, examples of 100% metering in many slums were shown to crack the hard nuts.

6.6.4 Ensuring DFT (IS 10500) quality of water at every consumer tap

It was easy to control the quality parameters in a few centralised production and treatment points. The real challenge was to ensure DFT quality in the tens of thousands of consumer taps after the water travels through several kilometres of underground pipes from the treatment point, criss-crossing several drainage and sewer lines en route.

The secret of ensuring DFT (IS 10500) quality of water at every consumer tap is not in the process or technology but in maintaining integrity in terms of leak-proof piped networks and deployment of smart water management to monitor the critical parameters. Furthermore, the Jal Sathis, who are trained and equipped with water quality-testing kits, carry out periodic testing at the consumer end which helps in monitoring the water quality on a regular basis. As a part of the water quality monitoring and surveillance protocol, state-of-the-art public–private partnerships water-testing laboratories have been set up at the regional level for regular testing of all required parameters from strategic sampling points. Mobile labs on wheels are also deployed for on-the-spot water quality testing. Real-time residual chlorine analysers at the far end of the network along with automatic chlorine dosing systems have been installed to ensure safe water even at the farthest consumer end.

People were sensitised about the potential water quality contamination in their overhead tanks and underground sumps. They were made to realise that unless the water is directly taken from the taps, bypassing intermittent storage, contamination cannot be prevented. Maintaining adequate pressure and availability of water 24 × 7 have demonstrated that household storage tanks are no longer required. Despite awareness creation, households usually take a few months with a wait-and-watch approach and finally abandon the household level storage tanks once trust and confidence on consistency and reliability of supply develop.

6.6.5 Tackling in-house knowledge gap and lack of executants with experience in 24 × 7

The absence of prior experience of the state in implementing 24 × 7 water supply and lack of sufficient precedence even across the country, with no implementation manual available, constituted one of the major challenges. Knowledge and prior experience gap existed at all levels starting from the top management down to plumbers at the field level.

To start with, a core team of key engineers of WATCO was sent to undergo a 4-day training programme at the ASCI, Hyderabad. Prof. V. Srinivasa Chary, Director at ASCI Hyderabad, has been training water utility engineers and advocating for 24 × 7 water supply in the country for several years. Inspired by his inputs during the training programme, the team was motivated by his passion for 24 × 7.

Further learning was gleaned from a successfully operating 24 × 7 system. Prof. Asit Biswas, a native of Odisha, a renowned global water expert and a recipient of the Stockholm Water Prize, conducted a customised 1-week training programme for six key members who were sent to the Singapore Water Academy of Public Utility Board, Singapore. The team was exposed to the nuances of the high-quality 24 × 7 water supply system operating in Singapore for decades. Although the team learnt the basics of 24 × 7 water supply technology including the design aspects, the Singapore training also exposed them to live operations. They learnt operational aspects and other equally important critical components of 24 × 7 such as consumer connect, customer relation, quality assurance, smart water management, water loss management and so on. Upon return, this team laid the roadmap for Odisha's 24 × 7 DFT water supply system, started implementing the programme and trained other engineers and staff of WATCO. In this way, the government was able to largely bridge the knowledge gap.

When the core team began preparing the tender document for engaging executing agencies, they realised it would be difficult to get executing agencies experienced in 24 × 7. Therefore, they decided to engage regular water supply contractors and train them on the job. This was not a one-time activity; engaging regular water supply contractors and training them on the job is continuing. This approach ensures adequate participation of bidders in the tendering process to avoid single tenders and facilitate competitive bidding, thus making the tendering successful, which is critical in the scaling-up process. At the same time, a larger pool of experienced workforce in the private sector is also being created.

A state-of-the-art 24×7 water supply system designed and built by eminent engineers may fail due to leaking house connections because of poor workmanship on the ground. One counter to that is training the plumbers. Local plumbers were trained and empanelled with WATCO for execution of leak-proof house connections. Soon the team realised the need to institutionalise the capacity building activities so that they were accessible in a continuous and sustainable manner. Taking a cue from the Singapore Water Academy, which is the capacity building and research institution of Singapore's Public Utility Board, the Odisha Water Academy (OWA) was established to function as a 'Centre of Excellence' for creating a trained workforce and continuous capacity building of staff. OWA (Now OUA – Odisha Urban Academy) imparts training and capacity building on water supply and sanitation sectors for all levels starting from operators to policymakers. Although Odisha created OUA to meet the state's own training requirements, it has now become an institution of national repute that has trained more than 1000 people including secretaries of urban development departments, chief engineers, private contractors, consultants and so on. OUA has been empanelled as a premier training institute by the Union Ministry of Housing and Urban Affairs (MoHUA). It has partnered with reputed institutions such as the National Institute of Urban Affairs (NIUA) and ASCI for conducting national- and international-level training programmes.

6.6.6 Augmenting human resources in consumer service delivery

Between 2015 and as of today, Odisha moved from 40% network coverage to 100% network coverage, from 0.3 million house connections to 1 million house connections. The scope of activities has increased manifold with corresponding human resources requirements. Increasing staff strength through large-scale government recruitments is extremely difficult. At the same time, these human resources were urgently needed to meet the increased requirements. With the human resources limitation, timely meter reading, bill delivery and revenue collection – which are essential to maintain the standard of services – were a challenge.

Odisha has had a strong women's SHG movement for the past two decades. It has actively engaged them to provide services under various programmes of the HUDD. Along with the Mission Shakti department, ULBs also form women SHGs and nurture them. Initially they were engaged in income-generating activities such as tailoring, making of snacks, papads and pickles, involving lower-level skill sets. The HUDD has recently embarked upon a journey of women empowerment by partnering with WSHGs in operating AAHAAR centres under the affordable meal programme, and for O&M of septage and solid waste management facilities involving financial and techno-managerial skill sets. Detection and regularising of illegal and unauthorised connections, mobilising new connections for achieving 100% house connection, providing doorstep delivery of services such as revenue collection are some of the activities which are difficult to accomplish with regular government staff. Furthermore, consumer connects, otherwise difficult to achieve, are possible through WSHG members who are chosen from the local community itself. Therefore, community partnership can be seen as the solution for several problems. Members from local WSHGs are appointed as Jal Sathis on a performance-linked incentive basis. They act as a bridge between WATCO and the consumers. The Jal Sathis are trained and entrusted with the responsibilities of meter reading, bill distribution, revenue collection at the doorstep with mobile point of sale (MPoS) machines. They are also provided with a water-testing kit and trained to carry out household-level water quality indicator tests. They are paid a specific amount for carrying out each such activity: 5% incentive for revenue collection, Rs 100 for facilitating a new domestic connection, Rs 200 for identifying and regularising an illegal connection, Rs 300 for a new commercial connection and Rs 20 for each water quality test.

The model of leveraging the power of community partnership with the onboarding of Jal Sathis as change agents has transformed the water supply service delivery and taken women empowerment to new heights. With this smart move, the government created a large brigade of workforce within a short span of time for improved service delivery, without having to resort to large-scale government staff recruitments. Jal Sathis have played a crucial role in urban Odisha achieving 100% piped water house connections in 105 out of 115 cities so far. The revenue collection efficiency has improved substantially; in Puri, for example, it increased to 97%. The smart innovation of Jal Sathis has become a game changer in the success of 24 × 7 DFT water supply service delivery in Odisha.

The community partnership model of engaging women for water service delivery has received several accolades at the national level. Inspired by Odisha's success of community partnership in the water sector, MoHUA, Government of India, made community partnership an essential ingredient of water supply service delivery in the national flagship programme AMRUT 2.0, encouraging other states to emulate the Odisha model.

6.6.7 Gaining public trust and confidence

Public water supply suffers from a legacy of poor water quality and lack of public trust. Generally, people do not trust the quality of water supplied by public authorities; rather, they trust water procured from private vendors, with even an unbranded water pouch carrying a higher trust value compared to the water supplied by public authorities. Furthermore, the public was apprehensive whether the government would be able to maintain continuous supply of water 24 × 7 without a break and with consistent pressure. Elimination of overhead tanks and sumps are essential to prevent contamination of water, and consumers in the state were advised to bypass their household overhead tank and sump and use water directly. As the public had already invested and built overhead tanks and sumps, they did not want to abandon them immediately, preferring to wait and watch.

As seeing is believing, it was decided to display key water quality parameters on a real-time basis on large LED display boards at high footfall locations across the cities. Apart from extensive information, education and communication (IEC) campaigns (such as 'Pure for Sure'), measures such as on-the-spot water quality testing by mobile labs on wheels and random quality testing of water in the consumer taps by Jal Sathis gradually helped build public trust and confidence.

To build public confidence and to encourage households to drink directly from the tap, an IEC campaign, 'Pure for Sure', was launched. Furthermore, real-time water quality data were displayed through digital display boards at key public locations. Execution of the programme in slums was challenging both from the point of view of convincing slum dwellers to opt for metred connections and operational difficulties given the layout and density in these settlements. Extensive IEC campaigns were undertaken wherein teams interacted with households to make them aware of the benefits of individual connections and good quality water, and the potential savings for the households. Execution of work in the narrow streets and alleys in slum settlements was undertaken during night hours to avoid any disruption to normal life.

6.6.8 Reduction of NRW

As per NIUA a study conducted by a third party, Odisha's NRW stood at around 54% during 2015. This meant 54% of the produced water was either getting lost on the way or not billed. According to the baseline data for Puri city, its NRW was 47%. It was understood from a very early stage of the DFT mission that unless NRW or water losses are controlled, 24 × 7 water supply services could not be sustained. One of the major risks in providing 24 × 7 water supply is that leakages in the pipeline result in 24 × 7 water loss, unlike for intermittent supply where the loss duration is limited to supply hours. A critical aspect of dealing with leakage control is to identify the point of leakage, as most of the piped network system is underground (mostly 1 m below the surface). Water by virtue of gravity tends to flow downwards when there is a leakage. At times it takes several months for the leaking water to become visible above the surface. Quick identification of the leaking points is critical to NRW reduction. This is one of the most challenging areas of 24 × 7 water supply relating to which new technologies are evolving every day across the globe. On the other hand, if water is supplied and not billed then 24 × 7 supply will encourage and result in irresponsible consumption.

At the design stage, each of the components of the 24 × 7 system is designed considering the population to be served and the per capita supply to be ensured with consistent pressure. With large-scale leakages and irresponsible consumption due to non-billing, demand will become erratic, making the supply system unsustainable, technically and financially.

Reduction of NRW has been given top priority right from the planning stage. Strategies were developed to address various aspects of NRW reduction and implemented on a mission mode. The ferrule point for a house connection – the

T joint connecting the street pipeline to the pipeline going to the house – has been identified as a potential leaking point due to unstable joints. Every defective ferrule point acts as a point of leakage and point of contamination. In a street pipeline of, say, 1000 m, there may be a ferrule point at every 5-m interval, making it 200 potential leakage and contamination points to tackle. Due to the enormity of the challenge, and especially when the urban sanitation policy of the DFT mission is to provide IS 10500 quality water at every tap, the team decided to replace all the existing ferrule joints with new ferrule joints with saddle and compression fittings to make them leak proof and contamination proof.

Furthermore, an industrial IoT system, embedded with multiple pressure- and flow-monitoring sensors installed at strategic locations across the pipe network, helps identify leakages in case there is an abrupt loss of pressure or flow in the pipe network, as they can be graphically seen online on a real-time basis. This helps quick identification and repair of leakages, resulting in substantial control of NRW.

An exclusive NRW cell with a dedicated team of engineers has been established at the WATCO head office to guide and monitor the implementation of NRW control measures in the cities. The NRW cell is equipped with state-of-the-art leak detection equipment. The team was rigorously trained on a month-long training programme to carry out NRW reduction activities effectively.

6.6.9 Quick complaint redressal

The manual mode of complaint logging by a consumer and assignment of staff for resolving the complaint used to take days to weeks, resulting in consumer dissatisfaction. Many times, the identification of the problem itself took a long time. In the absence of any consumer connect, the consumers often opted not to register complaints, preferring to live with the problem, partly because the complaint-logging process was cumbersome, and they were not hopeful of a response from the authorities. There was a need to simplify and streamline the complaint-logging and redressal process for improving consumer satisfaction and operational efficiency in service delivery.

To simplify and streamline the process, an interactive voice response system-based centralised customer care system was established for quick complaint logging, automatic assignment of complaint to the concerned engineer/crew for immediate action and online real-time tracking of redressal. For immediate response to complaints, leakage control and supply-related incident management, quick response teams (QRTs) on exclusive mobile vans are deployed with necessary tools and equipment. QRTs are available 24 × 7 for rushing to the incidence sites for quick redressal. The deployment of the QRTs also helps immensely in reduction of NRW due to quick repair of leakages. The complaint logging and redressal process along with the functioning of the QRTs is monitored on a real-time basis in the smart water management system.

The quick deployment of QRTs to the spot for complaint redressal has greatly enhanced customer satisfaction and has worked as a great confidence-building measure. Even if the QRT is unable to immediately resolve the issue in case of major repairs, the immediate arrival of the team to attend to the complaint provides huge comfort to the consumers.

6.6.10 Project monitoring

When 24 × 7 DFT water supply was promised to the public, it was essential that the promise was delivered in a time-bound manner. As the aspirations of the people were raised and a significant hype was already created around DFT, any delay in completing the project in time would hit the credibility of the government. The self-set ambitious timeline of the government required implementation of professional project management practices at all levels.

The HUDD water team was very conscious of the need for close monitoring and arresting of time slippages. For real-time project progress monitoring, a web application, Project Performance Monitoring System, was developed and deployed. The field functionaries update their item-wise progress data from the field on a real-time basis. This helps in quick identification and resolution of field challenges for timely completion. Review meetings were held every Saturday. From the very beginning of DFT, the Saturday meetings were institutionalised. The Saturday meetings are taken by CEO, WATCO, with key field officials. Starting from the inception of DFT since August 2019, not even a single Saturday has passed without a DFT review at the CEO level; such is the passion with which the mission is implemented and monitored.

6.6.11 Proactive community engagement

Active community engagement has been an integral part of the planning and implementation stages. At the very outset, ward-level water and sanitation committees were created to ensure community oversight during project planning and implementation. A total of 2015 ward committees were formed, and the same number of ward monitors (one for each committee) were appointed. The ward-level monitors were either members of SHGs, Biju Yuva Bahini[5] or local NGOs. The ward monitors were instrumental in facilitating and supporting the Jal Jogan Melas, monitoring service delivery at the local level, and coordinating timely redressal of citizen complaints and grievances.

As noted above, in the early stages of programme implementation, resistance was reported from some sections of the public. To address the resistance and apprehensions as well as to instil public confidence, awareness and motivation IEC activities such as one-to-one meetings and group discussions with communities involving their local leaders were organised. Such initiatives have helped in eliciting the support of local communities and households.

During project implementation, it was realised that community connect is vital to ensure mobilisation of households to avail piped water supply connections as soon as the pipeline networks are laid. The idea of Jal Sathis, members of WSHGs, as an improved version of water monitors was born. The Jal Sathis have filled the gap in community connect and are acting as a

[5] Biju Yuva Bahini is a youth engagement programme in Odisha introduced by the Department of Sports and Youth Services, Government of Odisha, in March 2018. It is aimed at channelising the energy of the youth towards nation building by inspiring volunteerism and active participation in various social action and community engagement activities.

bridge between the consumers and the PHEO. With appropriate orientation and training, the Jal Sathis have been providing consumer-friendly services as they hail from the local community and are more accountable to their own community.

Tapping diverse funds for water projects

The improvement of water supply at Greater Berhampur is a great example of the use of different sources of funds for the purpose. Of the total project cost of 3720 million, 400 million was a loan while another 400 million was a grant from Odisha Urban Infrastructure Development Fund, 1650 million were from AMRUT and 1270 million were made available under the state plan.

The development of water supply infrastructure in the four towns of Jharsuguda, Baripada, Rourkela and Sambalpur, costing 3180 million, was funded through Urban Infrastructure Development Scheme for Small and Medium Towns Scheme and from funds available under the state plan.

In Brajarajnagar and Belpahar, water supply revamp, implemented at a cost of 2280 million, was funded through the Development Management Fund (DMF) and National Thermal Power Corporation.

The water supply projects at Rajgangpur, Sundargarh and Biramitrapur costed 1260 million which was funded through the DMF.

On 18 December 2019, Government of Odisha launched the Jal Sathi initiative at an event organised to commence collection of water charges by Jal Sathis from the consumer's doorstep using an MPoS device. A standard operating procedure was also prepared detailing the roles and responsibilities of Jal Sathis and the defining principles of partnership to bring more clarity to the alliance between the PHEO and Jal Sathis. The Jal Sathis have been trained and mobilised for meter reading, bill distribution and revenue collection at the doorstep with MPoS machines. They are engaged for doorstep water service delivery of water connections, billing and collections and grievance redressal on a performance-linked incentive partnership. This has helped bridge the gap between citizens and the government, as well as improved gender parity in urban service delivery.

6.6.12 Formation of WATCO

WATCO was incorporated in 2015 as a wholly owned government company to carry out the functions of operation, maintenance and management of production, distribution and management of water supply and collection, and treatment and disposal of sewerage in Odisha. It is envisioned to deliver services of water supply and sewerage to citizens across the state in a sustainable, efficient and transparent manner.

WATCO is incorporated as a government company limited by guarantee and having a share capital of Rs 10 million divided into 100 000 shares of 100 each. The present structure of WATCO is: it is headed by a board of directors followed by the chairperson, vice chairperson and the chief executive officer. The chief operating officers in the three zones are responsible for their respective jurisdictions.

6.7 PEDAL TO THE FLOOR

The efforts for expansion, and monitoring and quality control across the water supply value chain have seen the total drinking water supply raised by 224 MLD. The state-wide average of 81 LPCD rose to 146 LPCD in 2018–2019, which is more than the national service-level benchmark of 135 LPCD. Also, by December 2020, Odisha managed to ensure that every single individual of its ULBs was supplied with water over the national benchmark.

Since the expansion began on a mission mode in the sector, Odisha has added almost 6000 km to its water supply network, a 111% increase that has taken the total size of the network from 5383 km in 2014–2015 to 11 325 km in 2019–2020. This doubling of the network has had the desired effect of tripling the number of households with piped water supply, which rose from 0.31 million in 2014–2015 to 1.05 million by the end of 2020 (Table 6.2).

Odisha's growth in urban water supply has not been just a tale of brute numbers. The improvement has been comprehensive, with quality growing alongside quantity, as proven by a study of 21 cities conducted by the BIS in 2019.[6] Bhubaneswar ranked the second in the study, with 9 out of its 10 water samples adhering to national standards.

Table 6.2 Progress of water supply in Odisha.

Parameters	2014–2015	2018–2019	2019–2023
Quantity of supply			
Quantity (mean)	81 LPCD	146 LPCD	All ULBs supplying more than 135 LPCD
Quantity (no. of ULBs)	<70 LPCD (49) 70–135 LPCD (54) >135 LPCD (9)	<70 LPCD (19) 70–135 LPCD (55) >135 LPCD (30)	More than 135 LPCD (115)
Coverage			
Fully covered wards	1187	1463	2032
Partially covered wards	701	530	
Uncovered wards	153	39	
Distribution network (km)	5383	9742	11 325
House connections	310 000	520 000	1 038 497
Population benefitted	2 million	6 million	7 million Mar 2024

Source: HUDD, Government of Odisha.

[6] https://bis.gov.in/wp-content/uploads/2019/11/Drinking_Water_20112019.pdf.

This marked improvement in water quality has also resulted in heartening statistics from the health sector in urban areas. The number of people infected with jaundice saw a steep fall between 2014 and 2018, from 3608 to 569. Consequently, and with other interventions, no deaths due to jaundice were recorded from 2016 to 2018. Overall, this improvement in the water quality has resulted in fewer outbreaks of disease, reducing costs incurred on health, especially among the urban poor.

The initiative has also sparked off an improvement in the quality of life of the urban dwellers, especially the residents of slum dwellers. Improved piped water supply has led to a reduction in time spent by slum households, especially women and adolescent girls, for water collection.

Improvements in both supply and quality of water from state-run utilities have meant that some typical expenses have been saved for households across the spectrum, more so for poorer ones. Estimates suggest that households would no longer need to cough up Rs 35 000 towards capital expenses for pumping, storage and treatment of water. In addition to this, they are also estimated to have saved between Rs 5000 and 10 000 on charges they would have paid to utilities for road cuts and restoration, which have been waived to incentivise house connections to the water supply network.

A smooth, well-organised improvement

Mrs Ahalya Das, aged 40, a resident of Maa Mangala Basti, Lumbini Vihar, Chandrasekharpur, Bhubaneshwar, is a self-employed vendor. A native of Khandapada, Nayagarh, she and her family have been living in Maa Mangala Basti for the past 20 years. Her family of five depends on vending chat and golgappa, and the family income is a meagre Rs 15 000 per month.

She shared that there were no disruptions in the water supply to their house and slum while the drinking water supply project was under implementation and the community water supply points continued to function. The family did not have to pay any upfront connection charges for their individual house connection as the piped water connections were provided under the Ujjala scheme. Her family also did not face any problems while seeking connections and the process was smooth.

Prior to the initiative, the community well was the only source of water; now Mrs Das has water supply up to her house. In addition, for drinking water there is a water ATM in the slum for which households are being provided with smart cards with a limit of 10 litres per household. She paid an initial amount of Rs 25 and now her family gets water free of cost. She used to spend about 30 min every day to collect water; now she can use that time for her work. She shares that she no longer needs to buy large storage vessels. Her one request was that the supply during summer months should be increased.

Urban Odisha's water journey has brought it several laurels. Odisha's all-round progress in water supply has been recognised at the national level. The state received the Housing and Urban Development Corporation award for best practices to improve the living environment in 2018–2019 for universal coverage of urban piped water supply. The state has secured the first position for successful implementation of AMRUT 1.0 across the country consecutively for 3 years – in 2018–2019, 2019–2020 and 2020–2021. The award for best civic agency instituted in memory of V. Ramachandran by Janaagraha City Governance Awards was received by PHEO, HUDD in 2020. The entire journey of 24 × 7 DFT quality water supply in Puri and in pilot zones in other cities such as Bhubaneswar, Cuttack and Berhampur has attracted global attention too. As a result, WATCO has received the prestigious 'Distinction of Global Water Leader Award, 2022' in the Global Water Summit held on 17 May 2022 in Madrid, Spain.

Taking note of the success of the DFT mission in Puri, the Government of India has mandated cities to provide drinking quality water on tap in its AMRUT 2.0 guidelines launched on 1 October 2021. Several components of the DFT mission find place in the AMRUT 2.0 guidelines – including community partnership, metering and reduction of NRW – as essential features.

Similarly, the Central Public Health Engineering Organization of MoHUA, Government of India, has published the success story of the DFT mission in Puri, which is inspiring water utilities across India to take up such projects. Water utilities from over 20 states have already visited Puri for knowledge exchange programmes, towards replicating the project in their home state. The community partnership model of engaging women for water service delivery has also received several accolades at the national level, encouraging other states to adopt similar community-led models.

As the next phase of progress in the water sector, the state is set to expand its DFT mission to cover 20 ULBs. It is fitting that Puri, with its famed Jagannath Temple and chariot festival (the origin of the word 'juggernaut') – a town that holds a special place in Odisha's identity – was the first city to be fully covered by DFT, setting the Sujal mission juggernaut (pun intended) rolling.

The DFT journey started in August 2019 with the sub-pilot in Ishaneswar Basti with a population of 2000, then moved to 12 pilot zones in two cities, catering a population of 0.19 million, to a population of 0.25 million covering the entire city of Puri. The mission has now moved to cover 24 cities with a cumulative population of 2.9 million, as of March 2024.

Odisha has committed to a holistic and integrated approach to water and has outlined intentions to move towards a coordinated approach to the management of the different components of the water supply value chain. The aim is to reduce the water footprint of urban areas and strive for greater sustainability and resilience. There is also greater awareness of and intent to shore up water utilities in terms of self-sustainability. In essence, by addressing every aspect of the water supply value chain, Odisha has shown its understanding of water's criticality to human lives, and the role it can play in aiding or impeding urban expansion.

Water crisis

Source: HUDD, GoO.

ESR, Masanichandi, Puri

Source: HUDD, GoO.

Jal Sathi

Source: HUDD, GoO.

Digital display board at AG Square 1

Source: HUDD, GoO.

doi: 10.2166/9781789064926_0101

Chapter 7

Aahaar: an affordable meal programme

ABSTRACT

This chapter documents Aahaar – an affordable meal programme being implemented by the Department of Housing and Urban Development, Government of Odisha and explains how the initiative aims at honouring the obligations of the state with respect to the right to food as well as the UN sustainable development goals relating to 'zero hunger'.

Keywords: Food insecurity, zero hunger, nutrition

7.1 HUNGER AS A CONTEXT

There is no city on earth where hunger does not exist. No matter how high the level of economic or social development of an area, there always remains a section of people who do not have access to two meals a day. Food insecurity is the critical factor that keeps people locked in a vicious cycle of poverty, wrecking any chances of upward mobility, while leaving them deeply vulnerable to even basic variations in the local economy.

Hunger and food insecurity could afflict a wide range of humanity: people from disadvantaged communities, the unemployed, students, patients and caregivers, commuters from remote villages coming to cities daily for work or higher education, people living in homeless shelters, senior citizens living alone or without incomes, those with physical or mental disabilities, households led by women or older people raising orphaned children and so on. What is worse is that while hunger is an immediate and pressing issue for many, the long-term impacts on nutrition could leave them stranded in even more vulnerable predicaments.

7.2 FOOD AS A RIGHT

Hunger as a human rights issue is derived from Article 25 of the Universal Declaration of Human Rights (UDHR), and Article 11 of the International Covenant on Economic, Social and Cultural Rights (ICESCR). Part of these visionary documents is an undertaking by ratifying states to take steps to address the question of hunger. India is a signatory to both UDHR and ICESCR. Zero hunger is also a global goal included in sustainable development goal (SDG) 2 which is to be achieved by 2030.

The recognition of food as a right then places some obligations on governments to not just prevent hunger, but to take proactive steps to guard against it in the future. (See boxes below.)

Right to food: obligations of the state

Respect: To not undertake any measures that arbitrarily deprive people of their right to food.
Protect: To enforce appropriate laws and undertake relevant measures to prevent third parties, including individuals and corporations, from violating the right to food of others.
Fulfil (facilitate and provide): To proactively engage in activities intended to strengthen people's access to and utilisation of resources so as to facilitate their ability to feed themselves. Whenever an individual or group is unable to enjoy the right to adequate food for reasons beyond their control, states have the obligation to fulfil this right directly.

Sources: UDHR and ICESCR.

Sustainable development goal 2 – need and targets for 2030

SDG 2: End hunger

- To end hunger, achieve food security and improved nutrition and promote sustainable agriculture.

Need

- Extreme hunger and malnutrition remain barriers to sustainable development.
- They create a trap that isn't easy for people to break.
- Hunger and malnutrition mean less productive individuals, who are more prone to disease and thus often unable to earn more and improve their livelihoods.

Targets

- *Target 2.1*: By 2030, end hunger and ensure access by all people, in particular the poor and people in vulnerable situations, including infants, to safe, nutritious and sufficient food all year round.
- *Target 2.2*: By 2030, end all forms of malnutrition, including achieving, by 2025, the internationally agreed targets on stunting and wasting in children under 5 years of age, and address the nutritional needs of adolescent girls, pregnant and lactating women and older persons.

About 11% of the world population – which is a massive 820 million people – are in hunger, according to the State of Food Security and Nutrition in the World 2019 report. Furthermore, 26% of the world population, or 2 billion people, experience moderate-to-severe food insecurity. India ranks 102 out of the 117 countries on the global hunger Index 2019. Its score of 30.3 points on this index classifies it as having a level of hunger that is 'serious'. The Indian state hunger index ratio termed Odisha's hunger problem 'alarming'.[1] The state was sixth from the bottom.

7.3 AAHAAR: APPROACH AND DESIGN

Aahaar was conceptualised as a programme that would enable the Government of Odisha to fulfil its obligation to provide food to people who are unable to realise the right to food on their own due to economic and other constraints. The Housing and Urban Development Department designed an affordable subsidised cooked-food programme for the underprivileged and named it Aahaar, which literally means 'food' in Odia, the local language. The work of preparing the blueprint of the scheme was initiated in December 2014 after studying existing successful affordable meal programmes operational in other states – such as the Amma Unavagam in Tamil Nadu and the Ram Roti Yojana in Madhya Pradesh.

Experience of implementing Aahaar

In 2015, Department of Housing and Urban Development (HUDD), Government of Odisha, initiated Aahaar programme with five urban local bodies (ULBs). It eventually expanded across 115 ULBs in Odisha.

Within a span of 90 working days, the development of the programme through iterative consultation was completed. This included designing training modules, arranging exposure visits for key government district and ULB officials to Aahaar centres and kitchens run by the Akshaya

[1] https://www.globalhungerindex.org/pdf/en/2019.pdf.

Patra Foundation, and building the capacity of implementing partners. Additionally, reviews of collectors and nodal officers of the Aahaar programme were undertaken, benchmark indicators for the programme were developed, Aahaar centres were graded and the quality and quantity of food were standardised, followed by standardisation of the entire process of the scheme.

The infrastructure of the programme encompassed Aahaar centres, kitchens equipped with all necessary appliances, inclusive access and seating arrangements for children, women, the elderly and people with disabilities, as well as vehicles to transport cooked food, all branded under the Aahaar programme.

Active involvement in the planning, implementation and expansion of the programme included testing the nutritional value of the supplied food and periodically assessing people's needs to further expand the programme. The programme aimed to bring equity and transformation for the poorest in society. A memorable photograph from the launch depicted a very poor child sitting next to the District Collector at a centre, sharing an Aahaar meal. This image encapsulated the purpose of the programme: bringing equity, eliminating hunger and showing care and sensitivity to the most marginalised and needy people. The goal was to ensure that no child, or any person, should starve at any point in life.

– Biraja Kabi Satapathy, UNICEF India

The programme was designed over a period of 90 days between January and March 2015. During this period, HUDD prepared the broad contours of the programme, which were deliberated upon and subsequently granted an in-principal approval by the state government; it was then designed and prepared for launch. On 1 April 2015, on Utkal Diwas (Odisha State Formation Day), Government of Odisha, launched the programme by serving the first meal at an Aahaar Kendra (outlet) located near a hospital in Bhubaneswar, the capital city of Odisha.

Amma Unavagam – mass affordable meal programme of the Government of Tamil Nadu

Amma Unavagam, which translates as 'mother's canteen', was launched in Tamil Nadu in 2013 by the late J. Jayalalitha's government. Amma Unavagams were meant to provide wholesome food at subsidised rates. Meals were priced at Rs 1, 3 and 5 and included *idli*, *pongal*, pre-mixed rice and chapatis with complimentary dal. Although the canteens were being financially supported by the state government, they were being operated by women's self-help groups (WSHGs).

7.4 CORNERSTONES OF THE PROGRAMME

Aahaar, which entails state-led provision of affordable meals for the poor, was pitted against the popular misconceptions about mass affordable meal programmes, which include that such programmes are populist and are implemented only for political gains, that they discourage people from working hard and earning a living and that they misuse the taxpayer's hard-earned money. Odisha's Aahaar was conceptualised as a human rights initiative by the government that would help the needy and the disadvantaged realise their right to adequate food and nutrition. The main objective of the programme is to provide affordable cooked meals to people who do not, due to their physical, social or economic conditions, have access to food.

The programme has five cornerstones including (1) no pre-defined beneficiary group, ensuring it is open and available to all; (2) affordable cooked meals; (3) use of a standard measure per serving; (4) accessible location of Aahaar Kendras and (5) superior quality and inclusive infrastructure with adherence to strict hygiene standards.

(a) *No pre-defined beneficiary group*: Aahaar is a self-targeted programme with no pre-defined beneficiary group, implying that any person can eat at Aahaar Kendras. There is no documentation or certification required to avail the service, a provision which was incorporated to avoid any middlemen or harassment to the deserving beneficiary. The fact that there is no pre-defined beneficiary has ensured that the programme has universal access. The following were, however, considered the target audiences: unorganised workers (including hawkers, rickshaw pullers and construction workers), patients and their attendants and any person who at that point of time felt that he or she was poor and needy. In addition to affordable food availability, the programme provides psychological support to patients and caregivers or attendants and poor students studying and appearing for various competitive exams so that they can survive in cities and towns with a small amount of money for food.

(b) *Affordable cooked meals*: A key aim of the programme is to ensure affordable cooked staple meals, consisting of cooked rice and *dalma*[2] *with* pickles, with each meal priced at only Rs 5.

(c) *Standard measure per serving:* The programme has adopted a standard measure per serving. Every person is served through serving mugs of 1 litre capacity for cooked rice and half litre capacity for *dalma*. This ensures that everyone visiting the Aahaar Kendras is served the same quantity of food – a feature that takes care of any prejudices that may arise on the part of the staff serving the meals. However, the option exists for the beneficiary to seek additional quantity or lesser quantity as per need.

---

[2] Dalma is a traditional recipe of Odisha. It is made with split chickpeas cooked along with raw papaya, pumpkin, potatoes, other vegetables and spices.

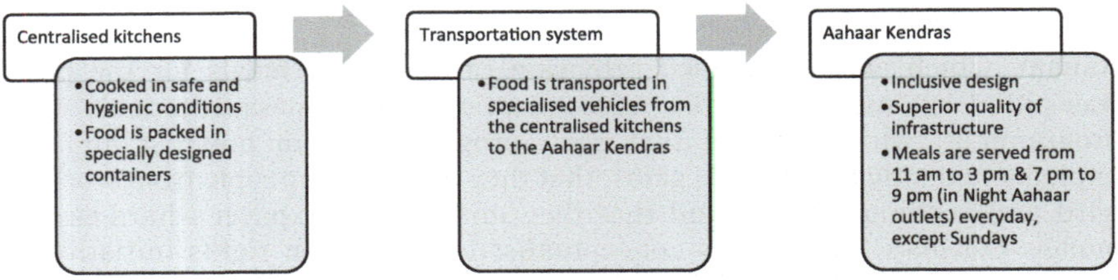

Figure 7.1 Programme implementation structure. *Source*: Authors' interpretation.

(d) *Accessible location of Aahaar Kendras*: Aahaar Kendras are located at places of public congregation. These are central locations that have high footfalls – hospitals, bus stands, railway stations, government offices, marketplaces and so on – making the centres highly accessible.

(e) *Superior quality, inclusive infrastructure along with adherence to strict hygiene standards*: The Aahaar Kendras have been designed to be inclusive, addressing the varied needs of the aged, infirm, differently abled persons as well as women and children. The infrastructure that has been created is of superior quality and is based on the philosophy that an intervention for the poor need not be of poor quality. In addition, along the entire value chain safe hygiene practices have been institutionalised and robust systems have been put in place to ensure adherence to these. There are also systems that allow regular monitoring to ensure compliance with safety and hygiene practices.

7.5 AAHAAR'S VALUE CHAIN

The Aahaar programme has three crucial components: the kitchens where food is cooked under safe and hygienic conditions; a dedicated transportation system with Aahaar branding (if the kitchen is not attached to the Aahaar Kendra) for transporting the cooked food, packed in specially designed containers, from the kitchens to the Aahaar outlets and the Aahaar outlets, where the meals are served from 11 am to 3 pm during the day and 7 pm to 9 pm at night every day, except for Sundays. Sunday is used for general maintenance, cleaning of the centres, kitchens and transport system as is shown in Figure 7.1.

7.6 PROGRAMME IMPLEMENTATION

The programme was implemented in a phase-wise manner. The first phase was launched on 1 April 2015 in five municipal corporations – Bhubaneswar, Cuttack, Berhampur, Sambalpur and Rourkela – with 21 Aahaar outlets. After the successful implementation of the first phase and on the programme's first anniversary on 1 April 2016, it was expanded to cover 74 towns across 30 districts through 111 Aahaar outlets.

The next phase of the programme was launched a year and a half later, on 1 October 2017, expanding the coverage of the programme to 85 towns and the total number of Aahaar outlets to 124. Furthermore, in February 2018 Aahaar

Table 7.1 Phase-wise implementation of Aahaar: number of ULBs covered and number of Aahaar outlets.

Date of Launch	Number of Local Bodies Covered	Cumulative Number of Aahaar Kendras
1 April 2015	Five municipal corporations	21
1 April 2016	Expansion to 30 districts, 74 towns	111
1 October 2017	85 towns	124
11 February 2018	114 towns	158
31 December 2018		167 (including 56 night Aahaar Kendras)

Source: Authors' compilation based on stakeholders' interview.

was extended to cover all 115 ULBs in the state and the number of Aahaar outlets increased to 158. Later that year, night Aahaar outlets were launched at locations where food is needed at nighttime including hospitals, bus stands and railway stations. A total of 56 such outlets were launched on 31 December 2018.

Currently, there are 167 Aahaar outlets present across all 115 ULBs in the state; 58 of these are day-and-night Aahaar outlets. The Aahaar outlets are supported by 38 centralised kitchens, 85 specialised food-transporting vehicles and approximately 2000 workers. Around 0.1 million (100 000) meals are served daily except on Sundays, and it is estimated that since the commencement of the programme (1 April 2015) around 135 million meals have been served as shown in Table 7.1.

7.7 IMPLEMENTATION ARRANGEMENTS: PARTNERS AND THEIR ROLES

The programme is being implemented through a partnership between the state government, Mission Shakti WSHGs and the panel of non-governmental organisations (NGOs). The Government of Odisha's nodal department for the programme is HUDD. The need for credible partners for the successful implementation of the programme was realised right at the beginning. HUDD was partnered with 17 implementing agencies which include foundations, WSHGs in various locations, community-based organisations, business associations, institutes and clubs – all not-for-profit organisations. The implementing agencies include Touchstone Foundation (a sister concern of Akshaya Patra Foundation), Manna Trust (a sister concern of Naandi Foundation), Jana Kalyana Parisad – Balasore, Neighbour – Bhadrak, Gram Uthan Kendra Pada, Abhijit Sahayog Samiti Trust – Paradeep, Pragati SHG – Bolangir, Maa Baisnabi SHG – Titlagarh, Maa Mangala SHG – Phulbani, Arati SHG – Malkangir, Nari Jagruti SHG – Balimela, Sonepur District Rice Miller Association, National Institute of Professional Studies – Anandpur and Rotary Club – Nayagarh, which were partnered with the programme initially.

With the objective of localising the operation and maintenance (O&M) and ensuring the participation of the local community in such urban initiatives for their empowerment and self-reliance, the government has decided to entrust end-to-end operation of the Aahaar programme to SHGs. Accordingly, O&M of all 167 Aahaar Kendras have been entrusted to 158 SHGs in 2022. The responsibility of the management of the Aahaar kitchens and transportation of cooked food is also being shifted from NGO-implementing agencies into Mission Shakti WSHGs. Currently, the implementing agencies Touchstone Foundation and Manna Trust are the only NGO-implementing agencies engaged in the programme to set up 56 new Aahaar kitchens and hand over their O&M – including cooking and transportation of meals – to the Mission Shakti SHGs and their federations.

In terms of the distribution of responsibilities, the state government has shouldered the complete responsibility of making the capital investments required for the creation of infrastructure, including Aahaar Kendras, centralised kitchens, transportation vehicles, vessels to cook and transport the food and equipment required to cook and clean. In the very nascent stages of the programme, the state government created Aahaar Kendras in existing vacant buildings; subsequently, the Aahaar Kendras have been housed in specially constructed buildings. The implementing partners have been tasked with the responsibility of operating the centralised kitchens and the transportation mechanism as well as running the Aahaar Kendras. They are also responsible for employing the required staff for these operations.

Each meal now costs Rs 23; while Rs 5 is collected by the implementing partners from the beneficiaries, the remaining Rs 18 is paid by the state government as subsidy. The main sources for covering this subsidy component are corporate social responsibility (CSR) funds, local donations and the Chief Minister's Relief Fund (CMRF). There are instances of people contributing the whole or partial costs of meals to their area's Aahaar Society on important days of their life such as birthdays, wedding anniversaries, relatives' death anniversaries and so on.

7.8 PROGRAMME INNOVATIONS

The programme has adopted several innovative practices. All facilities under the programme, including centralised kitchens and Aahaar Kendras, are ISO 9001:2015 certified for their quality management systems. Furthermore, Aahaar has been successful in implementing a crowd-funding model by leveraging funds from a variety of sources; this includes CSR funds, local donations/charity and CMRF, as mentioned above. A detailed standard operating procedure has been put in place for all functions along the value chain. There is a well-defined training programme for all Aahaar staff which is conducted by the implementing partners and HUDD. The focus of the training programme is on ensuring adoption of safety and hygiene practices as well as customer service standards.

Another innovation of the programme has been the institutions that have been created for its successful implementation. Each Municipal Corporation has a City Aahaar Society (CAS) and each district has a District Aahaar Society (DAS). The CAS and DAS federate at the state level as the Odisha State Aahaar Society (OSAS). The CAS, DAS and OSAS are registered as societies under the Society Registration Act, 1860, and have a governing body as well as an executive body headed by the Principal Secretary of HUDD. The governing body of OSAS is headed by the Chief Secretary, Government of Odisha, and the executive body is headed by the Commissioner-cum-Secretary of HUDD. Each DAS has the Collector-cum-District Magistrate as its chairperson. The societies can mobilise funds from donors including corporates, industries, traders, philanthropic organisations, individuals and so on.

Another innovative aspect of this programme is that all ministers and government officials have been advised to eat at the Aahaar outlets while they are on official tours in the state. This helps to bring visibility and focus to the programme and strengthen it, as such visits serve as surprise checks on its functioning.

7.9 AHEAD OF THE CURVE

Governments tend to resort to such interventions usually in response to crises. The Odisha's government's decision to tackle urban hunger without a trigger event is symptomatic of its proactive stance on welfare. The funding of the Aahaar initiative such that it is free of the strings of budgetary outlays is a key feature of the programme, allowing it to function independently and self-sufficiently.

Apart from their usual function, Aahaar Centres have also played a critical role in ensuring a certain amount of food security during the COVID-19 pandemic. More than 10 million meals were served during the pandemic, in 2020 and 2021, through takeaway and other safe modes for distribution of food. The centres have also served as food distribution points in the aftermath of cyclones, which are a frequent occurrence in Odisha.

Aahaar Centres have proved to be very popular and can be considered for expansion in all urban pockets of the state, apart from their extension to rural areas.

Baba Ambedkar Bus Terminal Bhubaneswar

Source: HUDD, GoO.

Netaji Bus Terminal Cuttack

Source: HUDD, GoO.

Netaji Bus Terminal Cuttack

Source: HUDD, GoO.

Netaji Bus Terminal Cuttack

Source: HUDD, GoO.

doi: 10.2166/9781789064926_0113

Chapter 8

Land rights for slum dwellers: Jaga mission phase 1

ABSTRACT

Recognising the significant contribution of slum dwellers to the city economies and growth, the Odisha government enacted Odisha Land Rights to Slum Dwellers Act, 2017 and expanded it with Jaga mission, aka Odisha Liveable Habitat Mission in 2018 to ensure inclusive development. This chapter on Jaga mission phase 1 details about the overall objectives, enabling environment and the detailed procedure for granting land rights to slum dwellers.

Keywords: Home-grown legislation, slum land rights, heritable but not transferable, urban scheme access, liveable habitat

8.1 CITIES AND SLUMS: A PRIMORDIAL CONNECTION

Slums have been an inevitable part of the urban fabric for centuries and remain one of the biggest challenges in the global urbanisation process. They are a manifestation of the inequities in urban life and an important sign of the inability of a city to cope with the challenge of providing necessities to incoming migrants.

Slums by nature are poverty traps and slum dwellers are unable to escape from this vicious cycle for generations, making it difficult for them to improve their living conditions. The absence of tenure security, lack of basic civic services, health facilities and economic opportunities, exclusion from connectivity, and most importantly, apathy of the government functionaries – all these factors accentuate poverty and make upward mobility and the gaining of social capital an uphill task.

However, substantive improvement in living conditions has the potential to mitigate some of these effects on economically vulnerable populations. Apart from fostering an urban framework that is inclusive, ensuring that development reaches these neglected pockets of the city, the critical need in this regard often boils down to the provision of formalised property rights.

8.1.1 Starting low, aiming high

Odisha is one of the least urbanised states in India. Its urban population is only 17%, well below the national average of 31% (Census 2011). However, the state has shown an upward movement in its urban count over the past couple of decades. And this trend is only expected to go up further.

To keep up with the pace and challenges of urbanisation, Odisha has demonstrated an increased focus on pro-poor reforms and interventions. Many programmes and interventions are directed towards the welfare of the 1.8 million slum dwellers across the state.

Recognising the critical contribution of slum dwellers in building and operating the cities, the Odisha government has decided to give land rights to slum dwellers under Jaga mission. The goals are to provide security to the vulnerable population in terms of their residence, and weave them into the very fabric of the city they live in.

Odisha's then Chief Minister on slum dwellers

My father had deep concern [for] those who lived in the basti. We also have to look deeper into the problems of unemployment. We are going to try for redressal; our duty is to fulfil my father's dream.

– Speech on Foundation Day of Biju Janata Dal (BJD), 26 December 1997

We have taken a historic decision today to grant land rights to the slum dwellers living in Municipal Councils and Notified Area Councils of the state and property rights to those living in Municipal Corporation areas. The cabinet today approved promulgation of two ordinances for assigning land rights and property rights to the identified slum dwellers for development, rehabilitation and upgradation of slums.

- Cabinet speech, 2017

Accordingly, Odisha enacted two ordinances on 10 August 2017 to grant land rights to eligible slum dwellers for redevelopment, rehabilitation and upgradation of slums across the state. The Odisha Land Rights to Slum Dwellers Ordinance, 2017, guarantees land rights to the urban poor households (HHs) in municipalities and notified area councils (NACs), while slum dwellers in five municipal corporations have been granted property rights under the Odisha Municipal Corporation (Amendment) Ordinance, 2017. The Odisha Land Rights to Slum Dwellers Act (OLR Act) was subsequently passed in the Odisha Legislative Assembly in September 2017.

8.1.2 Roots through rights

An increase in economic activities and increased economic opportunities contribute to the pull factor, leading to a rapid flow of migrant populations. These migrants get absorbed in various informal works in the cities as daily wage labourers, construction workers and so on. As the cities lack provisions for housing them, they usually end up in an existing slum, generally with the consent of the influential people of the slum, or form new slums, near the work site; they have no security of tenure and a precarious livelihood.

Although it is evident that residents of these slums are productive agents in the urban economy and contribute immensely to the development of the city, this population is often considered a 'burden on the city'. Instead, the Government of Odisha has decided to provide land rights and basic infrastructure to every slum dweller and this has led to the enactment of the landmark Odisha Land Rights to Slum Dwellers Act.

> Evicting slum dwellers or empowering them? We decided to empower them by providing land rights. We value people more than land.
>
> **-Chief Minister, Odisha**
> **- Cabinet speech, 2017**

The legislation stands out in the policy arena in India and globally today because it is bold and revolutionary within a democratic governance framework. This landmark decision was born out of the state's desire to transform the life of the slum dwellers rather than triggered by external funding or incentives.

Access to many schemes such as the Smart Cities Mission, Atal Mission for Rejuvenation and Urban Transformation (AMRUT), Pradhan Mantri Awas Yojana (PMAY), National Urban Livelihoods Mission (NULM) and Swachh Bharat Mission (SBM) has been limited for slum dwellers because of the informal nature of their dwelling. Their inability to produce proof of ownership or tenancy of their dwellings made them ineligible for financial assistance to make improvements to their homes. Furthermore, the informal nature of their dwelling arrangement meant that they were not consulted in any part of policymaking, and their localities were left out of the wider planning process. This Land Rights Act also helped in availing various benefits of other state and central government schemes which they might not otherwise be able to tap due to the absence of land rights or due to the informality of the settlements.

The journey of Odisha's Urban Land Rights Act

The importance of ensuring that the Odisha Land Rights to Slum Dwellers Act was fit for purpose cannot be overstated. At the Housing and Urban Development Department (HUDD), we engaged in numerous discussions with the Law Department of the Government of Odisha to update and reform various acts and legal provisions impacting urban development planning in the state. These discussions, spanning several years, enlightened us on the necessity of creating an act that was simple and primarily addressed broader framework issues, leaving finer details to be worked out during implementation. We also learned that attempting to revamp the existing Odisha Government Land Settlement Act (OGLSA), which had provisions for the transfer of pattas in urban areas since the 1960s but had not been implemented, was not appropriate for executing a government priority programme at scale. Therefore, the Odisha Land Rights to Slum Dwellers Act was drafted as a framework act under which rules could be notified from time to time to aid the implementation of the Jaga mission. This comprehensive framework-level act created an enabling environment and sparked new enthusiasm among urban local body (ULB) officers, district-level officers, revenue officers and citizens. Thus, the approach of creating a new simplified act with provisions for further detailing has been a key ingredient in the success of the Jaga mission.

- Shri. G. Mathi Vathanan

On 10 August 2017, the ordinance bill was approved by the Governor and the Act was notified in the state gazette. The bill was later placed before the Assembly in the following session of the assembly where it was passed with a majority of votes, despite a walk-out by the opposition members (who ostensibly welcomed such a law during debate hour!). Within a month of the gazette notification of the legislation, the relevant rules were formulated and notified to kick-start the implementation.

The Land Rights to Slum Dwellers Act, 2017, is an innovation in the field of tenure security, as it envisions giving statutory rights and entitlement – represented by land rights certificates (LRCs) – to slum dwellers over the land on which they live. These land rights are heritable but not transferable. The legislation thus gives conditional rights with protective clauses preventing sale and transfer.

A historic cabinet meeting for a momentous bill

It was the first cabinet meeting under the then CM since 2000 that was convened with just one agenda where the media briefing was done by the former CM himself. This showed the strong political will for the cause of slum dwellers.

- Shri. G. Mathi Vathanan

The drive from the top of HUDD combined with the strong backing for the cause from the Chief Minister's Office (CMO) had set the stage for a programme fraught with political risk, but it would come to be recognised internationally as a quantum leap in the right direction on tenure security.

8.2 APPROACH AND COVERAGE

8.2.1 Innovative approach

Although the HUDD was entrusted with the responsibility of implementing the act, the Revenue Department was not completely ignored; that department was marginally involved in the process. It was a balancing act: completely sidestepping it would hamper the implementation as it is the land custodian department and its functionaries – such as revenue inspectors, tahsildars and sub-collectors – would have to play a role in the entire land settlement process. But at the same time, they could not be given a prominent role due to the lack of commitment to the cause, and by training too, they are basically land custodians and believe in protecting the land: they would not be able to switch overnight to land giver mode.

Hence, the District Collector – now more a development and welfare functionary, from the viewpoint of the act, than a revenue and enforcement functionary – has been appointed the chairperson of the committee in each ULB vested with the authority of final approval of land rights, the Urban Area Slum Rehabilitation and Redevelopment Committee (UASRRC). The tahsildar has been made a member of the UASRRC. To make the LRC equivalent to the regular land patta issued by tahsildars in perception and to enhance the acceptance value of the patta by the community, the act assigns the tahsildar the role of authorised officer to sign the patta.

There was no precedent for implementing the key provisions of this act; hence a pilot was implemented in 54 slums across nine cities of the state covering 4257 HHs in two districts. This was performed to field test the processes and methodology developed by the state team with wider consultations and to develop standard operating procedures (SOPs) through field iterations before scaling up the implementation. This pilot implementation was initiated in one ULB of Puri district (Konark NAC) and eight ULBs in Ganjam district (Chatrapur, Hinjilicut, Khallikote, Chikiti, Gopalpur, Digapahandi, Polasara and Kabisuryanagar NACs). These two districts were selected based on leadership and track record/drive and coordination at the district level, and the ULBs were identified in consultation with the collectors considering the leadership at the city level. The pilot was designed to represent various types of slums across the state, to test replicability and scalability, leading to a selection of 54 such slums. It was crucial to test the implementation across at least two districts to maximise the learnings and to get diverse learnings, as district leadership plays a pivotal role in the whole process.

The objective of the pilot was to maximise the opportunities of experiencing the bottlenecks and to create a template covering all possible challenges, to make the process for scaling up easy. The importance of pilot testing for this kind of initiative is evident: to help in testing the validities of various assumptions made in the implementation process and methodology, and to reduce the scale of mistakes and degree of risk at the state-level implementation stage. For the pilot, operational guidelines were issued with certain assumptions, the field functionaries from the pilot areas were closely involved, and state coordinators were tagged to the pilot ULBs as mentors and as the communication channel between the state and the field. The state coordinators were responsible for transferring the policy implications to the field and seamlessly, for review.

Teams were formed at the city and district levels, which were then extensively trained by the state teams to undertake the pilot. It was vital to create a passionate team of foot soldiers with a mindset of problem-solving, not accepting roadblocks, keen on discovering new paths and marching on.

Expectedly, the experiences gained from the two districts were very diverse, which vindicated the rationale of selecting two districts for the pilot. The pilot was largely successful in Ganjam district but not so encouraging in Puri. Learnings as to what worked in Ganjam district and what did not work in Puri district helped in the preparation of the state-wide template for scaling up. One of the most crucial learnings the pilot brought out was the importance of the role of the District Collector.

Additionally, the pilot provided key learnings on the methodology of survey. In one ULB, the pilot undertook the conventional chain survey method through revenue functionaries, such as Amin, revenue inspector, tahsildar and an unmanned aerial survey using drones simultaneously in other ULBs. Thereafter, the process involved, time taken and difficulties encountered were analysed and compared: the conventional chain survey was found to be intensive, process oriented, time consuming and highly dependent on field functionaries. Moreover, it was estimated that it would take 33 years to complete the survey of all 2919 slums in 114 cities if the conventional method is used! However using drones, the survey can be completed much faster. (In fact, the survey of about 1697 slums at the scaling-up stage which was performed using drones could be completed in less than 6 months' time.) The pilot started around November 2017, and it was completed by April 2018.

Thereafter, a state-level function was held on 7 May 2018, at Chatrapur in Ganjam district, where the first set of LRCs was handed over to 2255 families from 54 slums in nine cities of two districts as part of the pilot initiative of the Odisha Land Rights to Slum Dwellers Act, 2017. The 'Odisha Liveable Habitat Mission – Jaga' was also launched during the event. Among the attendees were industrialist, philanthropist and Chairman of Tata Trusts, Mr Ratan N. Tata, and Lord Norman Foster, an internationally renowned and a Pulitzer awardee in architecture. Lord Foster in his address called the legislation to accord land rights to slum dwellers a path-breaking one, whereas Mr Tata termed it 'earth-shaking' and 'far-reaching'. The scaling up of the land rights programme in all the 109 municipalities and NACs was kick-started at this event as well, followed by the issuance of SOPs and guidelines.

> Odisha's initiative will have a ripple effect across the world. It will give a signal to the global community and aid their involvement in this kind of legislation (giving land rights and homes to the landless).
>
> **– Lord Norman Foster**
>
> (*Source: 'In historic move, Odisha gives land rights to 2000 slum dwellers', Times of India, 2018*)

One of the hallmarks of this project is that it is a highly collaborative one, involving multiple stakeholders and partners and association with one of the world's best agencies of architecture and planning, the Norman Foster Foundation. The state government, ULBs, non-governmental organisations (NGOs) and slum dwellers' associations (SDAs) are the main stakeholders in executing the act and implementing the Jaga mission. In addition, agencies such as Tata Trusts, Cadasta Foundation, Omidyar Network and SPARC Pvt. Ltd., Bhubaneshwar – apart from the Norman Foster Foundation – have contributed significantly through their resources, technology and knowledge sharing. There were 27 grassroots-level NGOs partnered with Government of Odisha for community mobilisation, continuous engagement with slum dwellers and collecting HH socio-economic data.

At the state level, the HUDD Principal Secretary, who is the Executive Chairman of the mission, calls coordination meetings at regular intervals with the key stakeholders to steer the programme. The District Collector drives the programme at the district and ULB levels. The programme management unit (PMU) of Jaga mission at the state and district levels provides support in this. The NGO partners are involved in activities such as community mobilisation, including continuous engagement with slum dwellers.

The project also uses advanced technology (unmanned aerial vehicles (UAVs) or drones) for capturing high-resolution images of the slum areas with their house footprints and for geo-fencing of the current slum boundaries. The execution of the project using high-resolution drone imagery provided a high degree of accuracy and transparency; coupled with field validation by the NGOs with the participation of the SDAs, the land rights settlement process was made strong and authentic. Following the imagery acquisition by drones, the technical agency integrated the imagery with geographical information system (GIS) data to create a community overview, supplemented with the data from HH surveys performed with the help of local NGOs. This led to the creation of an HH-wise database. A concurrent, updated and multi-layered database has thus been created for all urban areas.

One of the major interventions of the project has been the involvement of the community. The underlying philosophy of the government behind this mission – 'We value people more than the land' – has been embedded in the minds of those engaged in this programme. Hence the community is involved throughout the process as a partner. To ensure community ownership it was decided that the proposal for land settlement had to come from the communities. SDAs have been formed in each slum and involved in the entire process to ensure community ownership of the process. Women and community opinion makers are encouraged to play a lead role. The SDAs supported the NGOs in conducting door-to-door surveys and the numbering of dwelling units to generate HH data to be integrated with GIS maps. The SDAs are also responsible for resolving disputes (if any) among the slum residents and preparing a list of eligible beneficiaries and a new slum layout plan with the help of NGOs. The role of the ULB is limited to facilitating the process, to inculcate community ownership.

Another innovation is the Land Entitlement Certificate (LEC) distributed to the slum dwellers residing on untenable land, as an assurance for grant of land rights later subject to fulfilment of conditions as per the Odisha Land Rights to Slum Dwellers Act, 2017, and the Rules. An interim measure for giving statutory recognition, it can be used for availing other benefits including address proof. It covers HHs on untenable land or private land or slums on temple land, erstwhile princely states (Raja land) or forest land.

> Jaga mission is considered successful due to the participatory and active involvement of the community members, Slum Dwellers' Associations and ward officers. This makes it a unique people-centric programme where the SDAs are key executors with a distinctive monitoring role.
>
> **– Dillip Kumar Mohanty, Executive Officer**
>
> (*Source: Interview, 2022*)

8.2.2 Coverage

This act is applicable in slums or slum areas, defined as a compact settlement of at least 20 HHs with a collection of poorly built and temporary tenements, crowded together and with inadequate access to basic services. This can be on any land owned or acquired by the state government or its undertakings, municipal council or NAC, irrespective of the land ownership status. The act lays down provisions for the maximum area of land that can be settled in tenable as well as untenable slums.

The eligible slum dwellers are given heritable and non-transferable land rights for the area of actual occupancy of their dwellings, subject to a maximum ceiling of 45 m^2 in municipalities and 60 m^2 in NACs. The first 30 m^2 are settled totally free of cost if the HH belongs to the EWS category (EWS stands for the economically weaker section whose HH income is up to the limit as notified by the state government from time to time). The land in occupation over and above 30 m^2 is settled at a fraction (25%) of the benchmark value of the land. For a non-EWS slum family the whole land under actual occupation is settled at 50% of the benchmark value of the land.

To become eligible for receiving the LRC, the slum dweller must not have any other certificate issued under this act and must have been residing in the slum before or on 10 August 2017. Lands under forests, railways and so on and private land are presently considered untenable until they are made tenable through a conversion process wherever and whenever feasible.

Definitions as per the Odisha Land Rights for Slum Dwellers Act, 2017

'Tenable settlements' means settlements, as decided by the Urban Area Slum Redevelopment and Rehabilitation Committee (UASRRC), where the existence of human habitation does not entail undue risk to the safety, health or life of the residents, or those on sites that are not considered contrary to public interest or on land that is not required for any public or development purpose.

'Untenable settlements' means areas where the existence of human habitation entails undue risk to the safety or health or life of the inhabitants themselves or where habitation is considered by the UASRRC not to be in public interest.

8.2.3 Creating an enabling environment

Creation of an enabling environment for any project is of utmost importance for its effective implementation. Any project without a proper institutional framework or plan of implementation is liable to hit roadblocks during implementation. This was a challenge that the Government of Odisha foresaw, and hence it identified relevant stakeholders, trained them and made sure that every stakeholder was aware of the processes. Besides, Odisha's approach of taking up a pilot before full-fledged implementation helped the government understand the on-ground challenges and rectify them going forward. This properly planned and well-structured intervention resulted in its successful implementation.

As noted, the HUDD is collaborating with philanthropic organisations such as Tata Trusts at the state level on the project. Tata Trusts has signed a memorandum of understanding (MoU) with the Government of Odisha for providing technical support, including drone surveys, slum mapping and capacity building of different stakeholders. Additionally, the state government, through Tata Trusts, has established a partnership with the Norman Foster Foundation, Spain, for pilot implementation of the 'liveable habitat' programme and strategies for scale-up. It also collaborated with SPARC, Bhubaneshwar, for aerial imagery using drones in the pilot phase. Furthermore, for state-wide implementation, three technical agencies – SPARC, Transerve and Surbana Jurong – were engaged for drone mapping. The 30 districts of the state were allocated among the three technical agencies for the purpose.

About 27 NGOs were selected by the city governments at the ULB level and more than 700 field coordinators of the NGOs were trained by the state team to take up community mobilisation and the urban slum household area (USHA) survey at the slum level, including formation of SDAs and ground-level implementation. In addition, 100 'Jaga fellows' have been engaged through Tata Trusts to provide handholding support at the ULB level for coordination, capacity building of stakeholders and facilitating implementation. The Jaga fellowship programme created by Tata Trusts has helped hugely in scaling up, as the Jaga fellows have played a key role at the ULB level in driving the programme. Additionally, training has been provided to all the ULB-level executive officers and community organisers, concerned tahsildars, district officials and District Urban Development Authority team members on implementing the land rights act and Jaga mission components.

The Jaga mission aims to bring about the transformation of existing slums into liveable habitats, with the provision of all essential civic urban infrastructure, including roads, drainage, sanitation systems, access to toilets, LED streetlights, piped water supply to HHs, common work sheds, parks, playgrounds and so on. It is registered as a society under the HUDD, Government of Odisha, headed by the Chief Secretary, Odisha, as Chairman and Principal Secretary, HUDD, as Member Secretary. The objectives of the mission are:

- To grant land rights to all HHs living in slums in all municipalities and NACs in Odisha.
- To grant property rights to all HHs living in slums in all municipal corporations in Odisha.
- To transform the slums into liveable habitats with all necessary civic infrastructure and services at par with the better-off areas within the same ULB.
- To improve settlements by enabling HHs who have obtained land rights to improve their housing through public housing subsidies.
- To mainstream the poor through participatory planning and budgeting by enabling slum dwellers to participate in planning for their settlements.
- To leverage and converge various schemes/programmes/funding opportunities by strengthening collaboration among various departments, ULBs, NGOs, financial institutions, international agencies, trusts, communities and other stakeholders.

8.3 PROJECT COMPONENTS

The primary goal of the Odisha Land Rights to Slum Dwellers Act, 2017, is to provide eligible slum HHs with in-situ LRCs in the NAC areas and municipalities. The process involves preparing a land rights settlement proposal, grant of LRC by the UASRRC, and issue of LRC. To achieve the goal, the programme has been divided broadly into four components: (a) setting up institutional structures, (b) geospatial survey, (c) HH survey and (d) documentation.

Community engagement had been made as an integral part of all the above components to ensure a participatory approach.

8.3.1 Setting up institutional structures

The mission has adopted a decentralised approach with focus on community participation at all stages of implementation; most decision-making occurs at the ULB and slum levels. Although the HUDD houses the Jaga mission at the state level, an UASRRC has been constituted in each ULB. The UASRRC has been supplemented by a Scrutiny Sub-Committee (SSC) to verify the data collected through the survey on ground, randomly as required. This is to ensure that due processes have been followed and objections and suggestions received at the slum and ULB levels were duly considered. At the community level an SDA was set up to ensure community mobilisation and support the ULB as well as the agencies throughout the process, from conducting surveys to providing LRCs.

Although reviewing progress, it was observed that there were inordinate delays at the SSC level in some ULBs. The reason was that the SSC – which includes the tahsildar – adopted a conservative approach and started verifying every case by redoing the field survey process. It was observed that some of the field functionaries were so conservative and nervous that they did not want to approve anything or sign any document without personal verification.

8.3.2 Geospatial survey

Geo-fencing of the current boundaries of about 2000 slums using drones has been completed. It was decided that drones would be used to capture high-resolution images of the slums for marking their external boundaries, capturing the house footprints and detailed mapping of the slum itself. Considering the compact, congested and very high-density occupancy in slums, conventional survey techniques (total station, etc.) could have been inefficient, more time consuming, more expensive and excessively dependent on the field-based Revenue Department functionaries. To ensure mapping and linking of HH data of each slum HH without any human error, large-scale mapping (1:500 scale, 5–10 cm accuracy) through aerial photographs has been performed and embedded in GIS. After analysing various options (such as satellite imagery), aerial photography providing high-resolution images was judged the most appropriate. This high-resolution imagery helped in reducing conflict and giving a clear picture of the plots and the boundaries of slums.

8.3.3 USHA survey

The USHA survey has been facilitated by the SDA with community participation, which has brought transparency and openness into the process. The survey is a key step in the implementation process. The objective of the USHA survey is to fill up the application form and collect data on slum HHs for determining the eligibility and area to be settled. The NGO partners conducting the surveys filled up the application form including demographic and socio-economic status, the access to infrastructure of the dwelling unit and the land area occupied. Thereafter, they collected documentary evidences including proof of identity and residence on or before 10 August 2017 (cutoff date), photographs of the applicants, supporting documents to prove identity, address and income (such as the below poverty line card), proof of stay, scanned thumb impression of spouses or adult family members and so on. These documents, along with the survey responses, were used to determine the HH's eligibility for getting land rights and as input for calculating the area to be settled.

The effective blending of the participatory process with the use of modern technologies such as UAV-based photography coupled with the USHA survey has led to the successful implementation of the project of granting land rights, with zero disputes and zero litigation to date.

8.3.4 Preparing a final list of eligible beneficiaries

After the surveys were complete, the SDAs were provided with the base map and the area statement based on which the SDAs, facilitated by the NGOs, prepared the land settlement proposal. Each proposal includes lists of beneficiaries, area occupied, area to be settled, premium proposed to be collected and also HHs on objectionable land or land obstructing the access road for utilities (making the road inaccessible for fire tenders and ambulances); such HHs obstructing the right of way are not to be settled with land rights but are to be suitably relocated via a process facilitated by the concerned SDA and NGO. HHs on objectionable lands are not listed for providing LRCs unless they agree to shift back to or relocate within the slum. The SDA facilitates the discussions with the HH and once they agree they are cleared for LRC. The land settlement proposal is then displayed within the slum, with 7 days' time allotted for registering objections. After considering and resolving the objections and suggestions, the land settlement proposal is submitted to the UASRRC, and is then routed through the SSC notified for each ULB at the district level.

With the information collected through the survey, the SSC verifies the eligibility of the applicant and their spouse, scrutinising the validity of the documents submitted as proofs of identity and residence, type of dwelling unit and area occupied for settlement of land. Additionally, the SSC verifies that the premium for any additional land areas is in accordance with the economic status of the HH and benchmark value for the land area and checks the validity of the documents for HH income proof. After verification of all the information, the list is published at the ULB level for receiving objections within 15 days. On disposal of the objections, the final list is prepared by the SSC and placed before the UASRRC for approval.

8.4 PROJECT IMPLEMENTATION

8.4.1 Detailed steps in the allocation of land rights

The implementation of the Odisha Land Rights to Slum Dwellers Act, 2017, began with the pilot in the two districts of Ganjam and Puri. The land rights allocation process is explained in six broad steps in the process flow diagram shown in Figure 8.1. The specific activities and stakeholders involved in each step are described in detail in the paragraphs below.

- **Identification, training and capacity building of stakeholders:** Training and capacity building of different stakeholders involved in the mission were performed in phases. The first inception meetings were held at each corporation, which kicked off the process of project

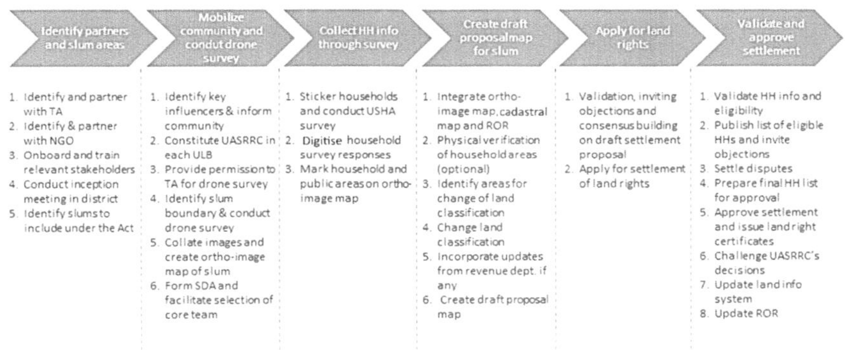

Figure 8.1 Process flow for granting land rights. *Source*: Adapted from Odisha: Land Rights to Slum Dwellers – Observations on detailed workflow process maps.

implementation and laid out the workplan, deliverables, support required for the smooth implementation of processes in the corporation. The training modules were prepared consisting of the act, rules, provisions, objectives, approaches, methodologies and roles and responsibilities of each stakeholder involved in the process. These modules maintained a uniformity in the work process without hampering the timelines set for completion of each implementation activity.

The second phase involved the identification and training of the survey agency that was responsible for enumerating the digital HH survey. Along with indoor presentation on roles, responsibilities, field work ethics and programme guidelines elaborate mock exercise were conducted in field. Demonstrations were conducted on how to identify houses from ortho-maps, how to sensitise the SDAs on the process, and how to conduct HH survey. These training sessions also included intricate discussions and live use of the mobile application used for conducting HH survey (Figure 8.2).

- **Clustering of slum boundaries:** The corporation team prepared the list of slums situated under the corporation area with the geo-coordinates, ward number, number of HHs (tentative) and submitted it to Jaga mission. They guided the technical team in identifying the tentative slum boundaries from the Google Earth and ORSAC maps and clustering of slums for smoothening the process of drone flying. The prior identification of tentative slum and slum boundaries helped in reducing the time spent by the technical agency in collecting the slum-level data.
- **Awareness and sensitisation of SDAs:** The SDA as an institution played an important role as a facilitator during the land rights implementation process. Meetings were conducted with each SDA in the slums and orientation was given to them on the act and rules, provisions, their roles and responsibilities and on addressal of conflict resolution, if any. The SDA helped the corporation, technical agency and survey agency

Sl No	House No	Beneficiary Name 1	Beneficiary Name 2	Occupied Area Remarks	Occupied Area(Sqft)	Village Name	Khata No	Rev Plot No	Kissam	RoR Tenant	RoR Area(Ac)	Area(Sqm)	Allotted Area(Ac.)	Alloted Area(Sqft)	Settlement Remarks
1	1	Reena Das	Bhararm Das	Ok	32.744	Hinjili S6(GA)	1539	1000	Bill Dofasli Ek	Sri Swapneswar Mahaprabhu Bije	0.198	3.042			
2	2	Jhunu Nayak	Kabi Nayak	Ok	29.493	Hinjili S6(GA)	1539	1000	Bill Dofasli Ek	Sri Swapneswar Mahaprabhu Bije	0.198	2.74			
3	3	Pinki Kumari Das	Sanjaya Das	Ok	105.755	Hinjili S6(GA)	1539	1000	Bill Dofasli Ek	Sri Swapneswar Mahaprabhu Bije	0.198	9.825			
4	4	Anita Das	Kiran Das	Ok	240.799	Hinjili S6(GA)	1539	1000	Bill Dofasli Ek	Sri Swapneswar Mahaprabhu Bije	0.198	22.371			
5	5	Jhili Das	Gandhi Das	Ok	296.869	Hinjili S6(GA)	1539	1000	Bill Dofasli Ek	Sri Swapneswar Mahaprabhu Bije	0.198	27.58			
6	6	Jhunu Das	Subash Das	Pay	325.468	Hinjili S6(GA)	1539	1000	Bill Dofasli Ek	Sri Swapneswar Mahaprabhu Bije	0.198	30.237			
7	7	Bishnu Das	Sankar Das	Ok	237.527	Hinjili S6(GA)	1539	1000	Bill Dofasli Ek	Sri Swapneswar Mahaprabhu Bije	0.198	22.067			
8	8	Sundari Das	Balaram Das	Ok	183.858	Hinjili S6(GA)	1539	1000	Bill Dofasli Ek	Sri Swapneswar Mahaprabhu Bije	0.198	17.081			
9	9	Chhaya Das	Malusu Das	Pay	338.288	Hinjili S6(GA)	1539	1000	Bill Dofasli Ek	Sri Swapneswar Mahaprabhu Bije	0.198	31.428			
10	10	Jhia Das	Kangali Das	Ok	254.857	Hinjili S6(GA)	1539	1000	Bill Dofasli Ek	Sri Swapneswar Mahaprabhu Bije	0.198	23.677			
11	11	saila Das	Bira Das	Ok	185.828	Hinjili S6(GA)	1539	1000	Bill Dofasli Ek	Sri Swapneswar Mahaprabhu Bije	0.198	17.264			
12	12	Urmila Das	Magati Das	Ok	274.878	Hinjili S6(GA)	1539	1000	Bill Dofasli Ek	Sri Swapneswar Mahaprabhu Bije	0.198	25.537			
13	13	Santosh Das	Maheswar Das	Ok	239.023	Hinjili S6(GA)	1548	2983	Kenal	Jala Sechana Bibhaga	0.54	22.206			
14	14	Namita Das	Daka Das	Ok	214.030	Hinjili S6(GA)	1548	2983	Kenal	Jala Sechana Bibhaga	0.54	19.884			
15	15	Saraswati Das	Sukuru Das	Ok	183.772	Hinjili S6(GA)	1548	2983	Kenal	Jala Sechana Bibhaga	0.54	17.073			

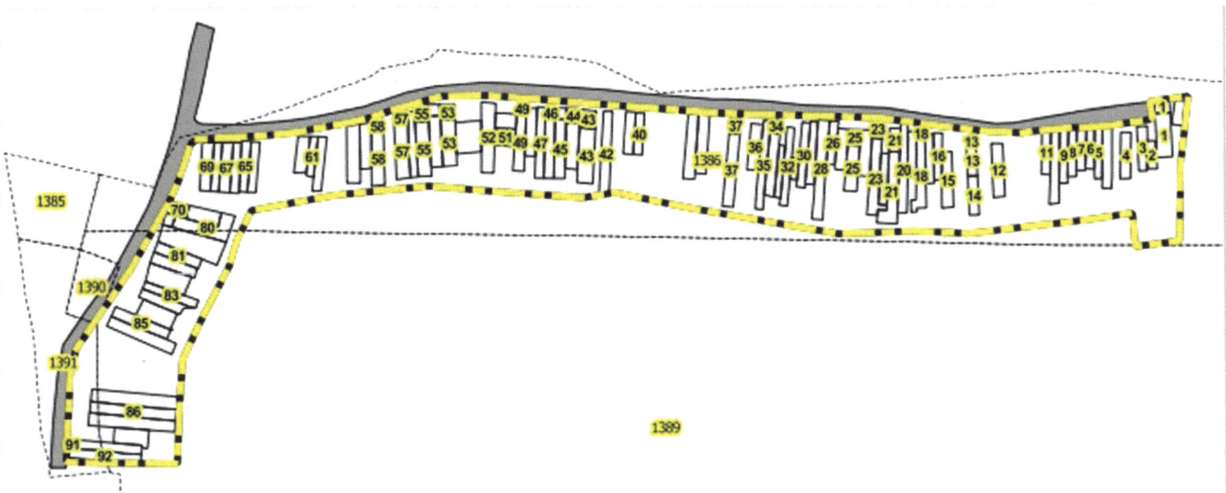

Figure 8.2 Sample of draft proposal map for slums. *Source*: Adapted from Odisha: Land Rights to Slum Dwellers – Observations on detailed workflow process maps.

in slum mobilisation, identification of slum boundaries during aerial survey, helping the survey agency in identification of HHs for stickering and digital HH survey, and maintenance of resolution for each activity to ensure transparency and accountability.

- **Aerial survey of HHs:** Drone survey is one of the vital components for land rights implementation as it brings inclusivity, accuracy and fastness in allocation of land rights to eligible beneficiaries. The visual template produced by this technology ensured larger public engagements, settlement of land issues with zero conflicts and ascertained the plots and boundaries. The technical agency used UAV-based high-resolution photography for the preparation of base maps for all slums in corporation. Corporation provided a road map to technical agency based on clustering of slums for ensuring speediness of the process. SDA meetings were conducted where slum habitants identified the extend of boundary. A consensus was taken from the slum residents, SDA members, ward officers, community organisers and corporators on finalising the slum boundary. After the drone flying, the technical agency processed the multiple aerial images captured into a single-stitched image to create a coherent ortho-image of slums, which was submitted to Jaga mission for starting the digital USHA survey.
- **Validation of ortho-maps:** After completion of drone survey, to ensure 100% accuracy of ortho-maps, the maps were validated by the SDA members and slum dwellers. The validation of maps was conducted to ensure that each HH in slums are covered in the generated maps. The SDA members along with community organisers from corporations and slum residents identified their HHs in maps and corrected the extend of the boundary, if any HHs or patch of land were missed out from them. The technical agencies were presented with the findings and wherever required, boundary revision of maps was performed or due re-fly of drone was conducted to retain accuracy in drone flying.
- **Stickering of HHs:** Stickering process provided a unique identifier to all HHs present in slums. A stickering format was developed by Jaga mission to ensure uniformity and standardisation in the process. The sticker included five codes with 14 digit number: district code (00), ULB code (000), ward number (00), slum code (000) and household number (0000). The stickers were pasted on each HH marking their inclusion in HH survey and for referencing the HHs for future visits for verification. The SDA members along with the slum residents guided the enumerators from survey agency in this process to ensure all families residing in slums are allotted with a unique identifier number.
- **Digital door-to-door HH survey:** The HH survey was conducted digitally to determine the eligibility of slum dwellers in getting LRC. The details collected through HH survey included socio-economic status, demographic details, infrastructure of the dwelling unit and chauhadi revenue details. Prior to the conduct of HH survey, on-field training was imparted to the enumerators from the survey agency to avoid errors of exclusion of beneficiaries. The SDAs were mobilised for administering

the process, facilitate their interaction with each HH and verify the authenticity of respondent. To ensure that HHs have required documents available with them during the survey, house-to-house dissemination of information on documents required, process and procedure was conducted. Along with that, steps were taken for miking announcements using Swachha sabaris (sanitation vehicles) for faster and wider coverage of the information. After the completion of the survey, the SDA members maintained a resolution in their register to validate and confirm on completion of the survey and coverage of all HHs present in the slum.

- **Marking HHs and public areas in ortho-maps:** The survey enumerators along with conducting digital HH survey also marked the ortho-image of slums for outlining the overall layout of slums. The final markings were used by the technical agency to calculate the area to be settled for each HH and plan the draft settlement map for each slum. The enumerators marked the HHs in maps as per stickered number, slum roads, access road of HH, public area and common area structures such as temple, pond, playground, open space, community hall and so on. Different coloured pens were used to mark the different structures present in the slums –for example, black to indicate HH number, red to indicate slum road, blue to indicate access road to the house and green for indicating public or common area. After preparing the final marked ortho-map the survey agency submitted them to technical agency to determine the layout for settlement and identify areas for shifting HH for relocation purpose, if required.
- **Creation of draft proposal settlement map:** The marked ortho-maps submitted by enumerators of survey agency was superimposed with the cadastral map and revenue records by the technical agency to create a draft proposal map. The draft proposal map was prepared by using the area statement (containing revenue and land details) and draft settlement map or tracing sheet maps that consisted of slum's layout. The draft proposal maps were used for validation and consensus building on area and type of settlement proposed for each HH. The technical agency after reconciling the data shared the draft proposal map with the corporation for validation and authentication of the revenue details generated in area statement and tracing sheet maps.
- **Support in revenue verification:** The corporation with the help of their revenue officials verified through the bhulekh (land records) – the revenue and land records generated in the area statements and physically and manually measured 5% of total generated data to authenticate and validate the draft proposal map. The revenue officials from the corporation after authentication, identified the tenable land where in-situ land rights could be allocated to the beneficiaries, and identified untenable land to pursue for classification change of Kissam or for rehabilitation process. The draft proposal map was used to prepare the list of draft eligible beneficiaries.
- **Recommendation by SSC:** SSC facilitated and ensured the implementation of rules, provisions, and guidelines of the act in the field without any deviation. It verified and validated the proposal of land rights for the eligible beneficiaries of slum and slum development from the corporation.

After scrutinising the HHs based on the land settlement and the eligibility from the documents submitted by the beneficiaries, the committee recommended for the preparation of the draft eligible beneficiary list to be approved in Slum Development and Rehabilitation Committee. A thorough exercise was conducted to weed out any malpractices observed in the process. Wherever inconsistency of data was found, the survey agency was instructed by the corporation to re-verify the data in the field. Information education and communication materials in the forms of posters, banners and making announcements were introduced to create awareness and sensitise slums on land rights process and prevent unscrupulous activities undertaken in the field, if any.

- **Publishing the list of draft eligible beneficiaries and inviting objections:** After the preparation of draft eligible beneficiary list, it was published in different locations – slums, District Collector's office, corporation office, tahsildar's office and ward office for inviting objections on the notice if any from the slum dwellers. This process is created for the slum dwellers to raise their queries, concerns and doubts if they feel that the recommendation made by the committee requires modification or change. The objection notification is published for 7 days, within which objections are raised. The corporations decided on ways to suggest appropriate action on the objections raised. If the dispute was unable to be resolved, then the matter was taken to the Slum Development and Rehabilitation Committee for final settlement of the dispute. After the completion of the 7 days, the final list of beneficiaries was recommended to the Slum Development and Rehabilitation Committee (SDRC) for final approval.
- **Approval by Slum Development and Rehabilitation Committee:** The recommended list of beneficiaries was approved by Slum Development and Rehabilitation Committee. The approval also included the land area allotted and the payment needed to be made for settlement by the beneficiary as per the provisions of the act. LRC is handed over to the beneficiaries upon compliance to the necessary provisions.

8.5 INSTITUTIONAL ARRANGEMENTS FOR SCRUTINISING AND SANCTIONING LRCS

The implementation of the Jaga mission programme on the ground adopted a decentralised approach, involving all stakeholders at the state, district, city, ward and community levels. The institutional structure at each tier of governance is detailed below (Figure 8.3).

8.5.1 At the state level

At the state level, the HUDD is the nodal agency for implementing the provisions of the act and Jaga mission. The Chief Secretary of the state is the chairperson of the governing body. The Principal Secretary, HUDD, is the chairperson of the executive committee, supported by the director, municipal administration, as the mission director.

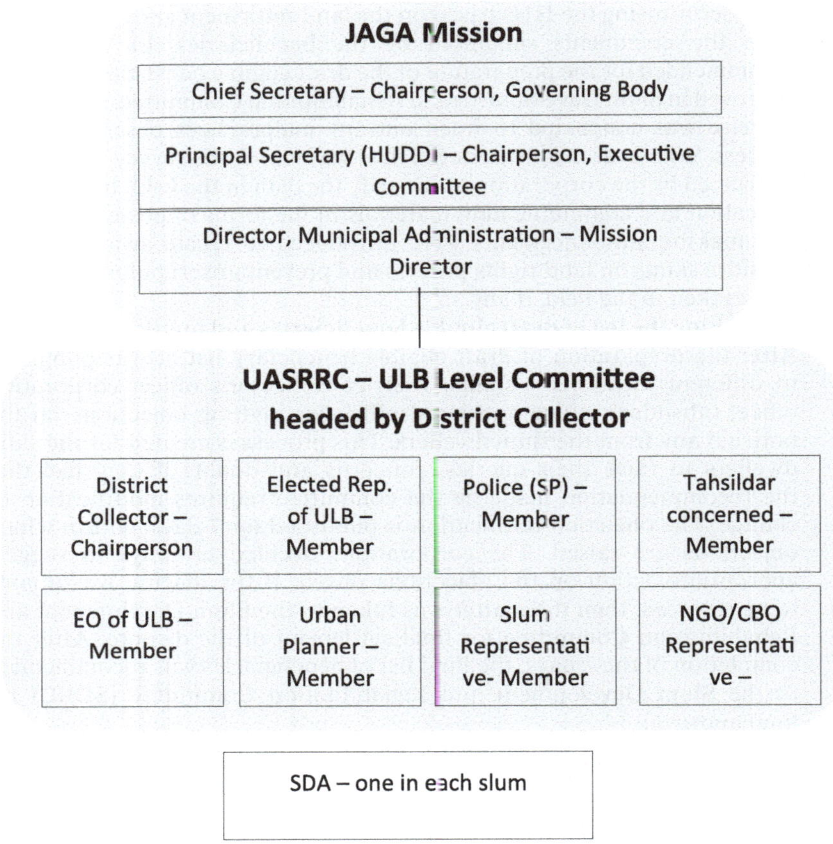

Figure 8.3 Institutional set up. *Source*: SCI-FI analysis.

8.5.2 At the ULB level

The UASRRC in each ULB is headed by the collector and constituted by the HUDD. The UASRRC is responsible for formulating the SSC and preparing the list of slum dwellers eligible for the issuance of land rights. The UASRRC typically consists of the following members:

- District Collector (head)
- An elected representative of the ULB nominated by the respective municipal council
- Superintendent of Police (or a representative not below the rank of Deputy Superintendent of Police (DSP) or sub-divisional police officer (SDPO))
- Tahsildar (of the corresponding ULB)
- Executive Officer (EO) of the ULB
- One urban planner (who is nominated by the District Collector)

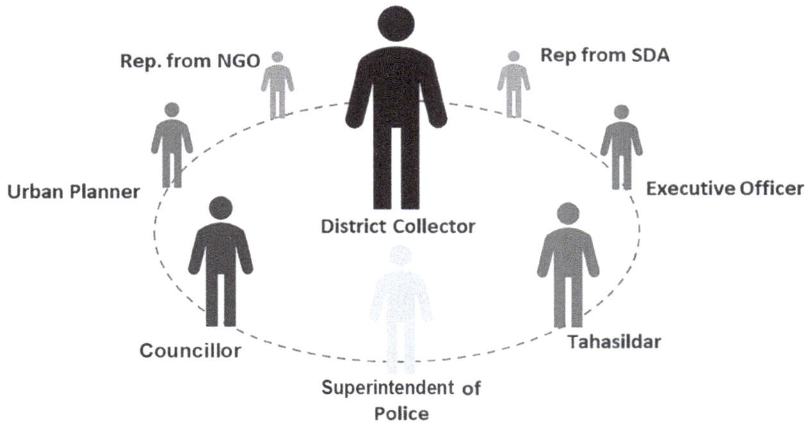

Figure 8.4 Composition of UASRRC. *Source*: Presentation made by PS, HUDD.

- One community-based organisation/slum dwellers' representative (who is nominated by the District Collector)
- One NGO representative (who is nominated by the District Collector) (Figure 8.4)

The SSC is responsible for verifying the applicant and HH information, and determining the eligibility and settlement details (area, layout, etc.). The Revenue Department, ULB and NGO support the SSC in the verification of land type and classification, HH details and layout and so on by providing additional clarifications, if required. In a few cases, members of the SSC go for physical verification of the HH/land. The SSC is constituted by senior district officials, representatives from the tehsil, ULB, NGO, Integrated Child Development Services organisations and health and family welfare organisations.

8.5.3 At the community level

The SDA is a community-based entity for each of the identified slums. The SDA comprises members from each HH in the slum. A core management team consisting of a president, secretary and treasurer is also elected for each SDA by the community. The SDA is a non-registered entity, acting on behalf of the community members and assisting the other stakeholders in the implementation of the act. In essence, the project envisages the SDA as an entity that truly represents the interests and preferences of the community. The SDA seeks to ensure that implementation on the ground involves active participation by the community. The key roles and responsibilities of the SDA include:

- Assisting the NGO in community mobilisation and USHA survey, and ensuring uniform community awareness and buy-in.

- Initial validation and building consensus on the draft slum settlement proposal provided by the technical agency. The draft settlement proposal details the HH layout and area to be settled for each HH, based on which the SDA applies for land rights to the UASRRC.
- Applying for land rights to the UASRRC.

8.6 BUDGET AND FINANCING

The HUDD is implementing the programme by utilising the existing human resources at the state, district and ULB levels. As their establishment cost is already provided and met from the annual state budget, no additional budget provision is required.

As discussed, Tata Trusts, a leading philanthropic agency, has been roped in as a leading partner to support the programme implementation. A state PMU with technical and managerial expertise has been set up to drive the programme with funding support from Tata Trusts. Tata Trusts is also supporting the engagement of about 100 fellows (known as the Jaga fellowship of young professionals) to support the implementation in the field. Tata Trusts has also brought in support from Omidyar Network and Cadasta Foundation for technological support such as drone surveys, slum mapping, online data collection and data digitisation.

The state government has provided a token grant of Rs 1 billion in the 2018–2019 and 2019–2020 budgets for mission implementation. Apart from this, the convergence of schemes and funds from central, state and other sources are ensured to fund the infrastructure and other services.

As a special gesture to salute the nobility of this programme, the Norman Foster Foundation has signed an MoU with Tata Trusts to support the design, development and execution of the in-situ upgradation of the slums and new habitats.

8.7 LINKAGES WITH ONGOING PROGRAMMES/MISSIONS

The Odisha Land Rights to Slum Dwellers Act, 2017, entitles the urban poor to rights on the land parcels they had been residing on, depending on the tenability. The dissemination of land rights to the urban poor, after the commencement of the Jaga mission, enhanced the potential beneficiary base in the state for leveraging available beneficiary-led individual construction housing subsidies under AWAAS (Odisha Urban Housing Mission). AWAAS had been launched on 11 October 2015, aligned with the national housing scheme, the Pradhan Mantri Awas Yojana (PMAY), in recognition of the need for an effective and efficient institutional mechanism for achieving the objective of 'Housing for All' in urban areas.

8.8 ESSENTIAL INGREDIENTS, GOOD RESULTS, FURTHER EXPANSION

The roll-out of the programme has created a bank of reliable data that has ensured the integrity of the process of providing LRCs and LECs. Odisha is now the only state in India to have detailed maps of all its slum settlements. Such a detailed spatial dataset has enormous potential for the optimisation of future poverty alleviation and urban planning schemes of the government, apart from opening the door to effective research on the quality of life in slum settlements.

Within about 18 months of its launch, a total of 1697 slums had been surveyed in detail using drones and geospatial technology in the USHA surveys. The surveys across the state have covered more than 0.17 million HHs. High-resolution maps of all the slums have been prepared to ensure that the recorded residents benefit from the programme.

More than 85 000 HHs have been given LRCs and over 84 000 have received LECs. A suitable blend of the participatory process with technologies such as drones, GIS and cloud computing has led to an implementation process that has seen no disputes and zero litigation. The two certificates have also enabled more than 42 500 HHs to avail financial assistance under AWAAS, the state's urban housing programme, in alignment with PMAY.

The project has also been instrumental in setting up institutions that have already proven to be a success in community participation in development efforts. A total of 109 UASRRCs and 1697 SDAs have been established. The SDAs have already proved their effectiveness as force multipliers in the implementation of schemes; an example is the central role they were mobilised

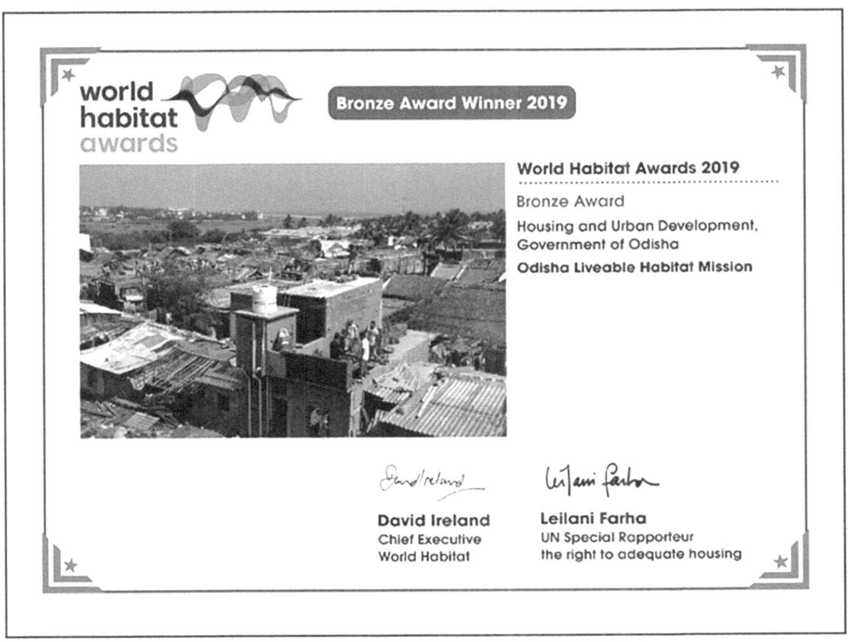

Figure 8.5 World Habitat Award 2019. *Source*: HUDD, Government of Odisha.

to play in the roll-out of the Urban Wage Employment Initiative at the height of the COVID-19 pandemic.

Recognition of the effectiveness and success of the Odisha Liveable Habitat Mission or Jaga mission has come from both home and abroad. It won a Bronze Award at the World Habitat Awards of 2019, where the panel specifically appreciated the speed with which the project was taken up along with its robust institutional and legislative framework. The Odisha government also won the 'India Geospatial Excellence Award' from Geosmart Media and Communication for technological innovation in empowering the urban poor (Figure 8.5).

Further expansion of the land rights programme has been started after another such historic step, on 5th November 2022, when the Government of Odisha flagged off aerial survey and implementation of land rights allocation in five municipal corporations of Odisha. Guided by The Odisha Municipal Corporation (Amendment) Act, 2022 and the rules made there under the following processes were followed to allocate land rights to 4.12 lakh population residing in 1222 slums of municipal corporations.

Although by no means a completed project, the Jaga mission has succeeded on two counts: it has largely achieved what it set out to do, and it laid the groundwork for further interventions. The second phase of the Jaga mission (discussed in Chapter 9) is in the process of being rolled out, its effectiveness bolstered by the depth of knowledge available as well as the involvement of community members. With its adaptability to further stacking of initiatives, the Jaga mission can safely be considered the foundation of Odisha's vision of slum-free and slum-proofed cities.

Slums before transformation

Source: HUDD, GoO.

Community mobilisation activities

Source: HUDD, GoO.

Drone survey of slums began from Konark. Pilot implementation began for nine ULBs

Source: HUDD, GoO.

Capacity building of SDA during land rights survey

Source: HUDD, GoO.

doi: 10.2166/9781789064926_0137

Chapter 9

Right to basic amenities and services to the slum dwellers: Jaga mission phase 2

ABSTRACT

In the second phase, towards transforming slums into liveable habitats and integrating the slums into the city fabric, the Biju Adarsh colony programme was undertaken. As part of upgradation, all basic infrastructure amenities and services have been provided in partnership with slum dwellers' association. In the untenable slums where in-situ land rights and slum upgradation are not permissible or desirable, new habitats are created in consultation with the slum communities and the relocation is voluntary. These new habitats also provide serviced housing plots to all the slum dwellers. With these efforts, all the slums get transformed into liveable habitats and the cities become slum-free. To sustain the slum-free status of the cities, the state is adopting slum-proofing strategies to pre-empt the formation of new slums.

Keywords: Slum upgradation, slum delisting, liveable habitats, basic infrastructure, community-based implementation, Biju Adarsh colony, slum-proofing, participatory needs assessment, i-Gap analysis

9.1 EVOLUTION = GROWTH

Any initiative, policy or programme starts with assumptions and an understanding of the problem. Although the problem can be observed and experienced, its full scope is rarely understood initially. This results in a tentative solution based on these early assumptions. However, as implementation progresses, the problem evolves, testing these assumptions, especially on a large scale. This variability necessitates constant refinement and adaptation. As the understanding deepens, solutions must be adjusted with new insights and corrections. This iterative process ensures the programme remains relevant, effective and successful. Static and rigid programmes often fail, while those designed for flexibility can adapt to real-world complexities.

The Jaga mission illustrates this need for continuous evolution. Initially focused on providing land rights through certificates, it soon became evident that tenure security alone was insufficient for true transformation. Recognising that land rights were merely a starting point, the mission expanded to include essential services such as safe water and infrastructure. This evolution shows a deeper understanding of empowerment and development. Successful programmes, especially government initiatives, must adopt this model of continuous improvement. By doing so, they stay effective and relevant, addressing the evolving needs and challenges of their communities. The Jaga mission's unique strength is its ability to continually upgrade, ensuring ongoing self-development and empowerment for slum dwellers.

The objective of improving the quality of life of slum dwellers in urban areas had been enshrined in the Odisha Land Rights to Slum Dwellers Act of 2017 and launch of the Jaga mission that year. In this framework, inclusive of development and integration into the city's economy and the fabric are as important as the creation of adequate infrastructure or the formalised provision of civic services such as water, sanitation or electricity.

9.2 LAND RIGHTS AND HABITAT UPGRADING AS A CONTINUUM

The first phase of the Jaga mission aimed to provide land rights to slum dwellers through the provision of land rights certificates (LRCs) and land entitlement certificates (LECs). These certificates have been issued to over 172 000 households (HHs), ensuring their tenure security. This was possible in slums where the government was able to secure the rights, which were termed tenable lands or in the process of converting the untenable to tenable by following due process of law.

In the first phase of the Jaga mission, the government committed to provide land rights for the slum dwellers; in the second phase, it has prioritised 'slum upgradation and delisting' through the creation of necessary basic civic infrastructure and bringing the tenable slums on par with the rest of the city. When the Odisha government started identifying the slums for intervention under the Jaga mission, they realised that about 50% of the slums were located on untenable lands, that there remained people at the margins of the cities occupying environmentally hazardous land or living under significant threat of eviction especially on railway or forest land. In-situ land rights could only be granted to slum HHs that exist on tenable government land. If the slum HHs exist on private land and/or untenable government land (e.g. environmentally hazardous lands, lands owned by the railways, forest or defence departments, lands reserved at some point in the past by the Revenue Department for waterbodies and grazing or lands owned by local kings, temple, etc.), then a range of measures – such as change of land classification, land exchange, slum re-densification and so on – need to be initiated to ensure optimum in-situ settlement of land. Only when all such measures are exhausted or not feasible only then would a relocation of slums to new 'liveable habitats' be considered, subject to the consensus of the slum dwellers through resolution of the slum dwellers' associations (SDAs). Hence, in the second phase, to secure the slum dwellers' 'right to the city' and adhere to the spirit of 'leave no one behind', the Government of Odisha has also initiated processes to address the untenability aspect and provide those occupying such lands with services and housing.

9.3 SAME GOAL, DIVERSIFIED APPROACH

In the second phase, the government adopted a two-pronged approach to address slums. First, it expanded the Jaga mission to take up slum upgradation in all places where LRCs and LECs had been issued and in the municipal corporation areas where Odisha Municipal Corporation (Amendment) Ordinance was passed initially in 2017 along with the Odisha Land Rights to Slum Dwellers Act. It was due to the realisation that property rights cannot be distributed given the huge resource requirements and time involved.

Even though land rights programme was not implemented in the municipal corporation area in the first phase, the state government went ahead and started implementing the habitat upgraded programme to ensure access to basic amenities and services to the slum dwellers and empower their associations. Thus, phase 2 of the Jaga mission was implemented in the corporation areas even before phase 1 implementation. Thus, delinking phases 1 and 2 makes them as independent and concurrent programmes.

The implementation of phase 2 marked the beginning of the government's collaboration with the SDAs in corporation areas. By establishing SDAs and executing the Mukhyamantri Karma Tatapara Abhiyan (MUKTA) yojana programme through them, the government gained invaluable experience and fostered a strong working relationship. This experience provided the confidence needed to introduce land rights in these corporation areas, acting as a precursor to the LRC programme. Without this foundational work, the assurance and strength to proceed with the LRC programme would not have been possible. The successful implementation of phase 2 alleviated the fear of risk and provided the necessary confidence to introduce the LRC programme in corporation areas. Essentially, the implementation of phase 2 laid the groundwork for the successful rollout of the LRC programme.

Understanding that raising the standard of services and conditions in the slums on par with the rest of the city is the way forward to integrating and mainstream slums with the city's spatial fabric. The upgradation of infrastructure, amenities and services in these slums would be followed by a delisting protocol by the city. The SDA would be converted into the Residents' Welfare Association (RWA) for the transformed habitats (Adarsh colonies).

The slum upgradation and delisting programme was extended to all municipal corporations, municipalities and notified area councils (NACs) in Odisha, and was limited to tenable slums or slums that can be made tenable in due course of time through negotiation, land exchange, land reclassification, re-densification and so on.

Universalising citywide access to services to integrate slums into the city

I had the opportunity to have detailed discussions on the Jaga mission – its approach and its successes – with the Additional Chief Secretary during a study visit to South Africa in 2018. During our long waits at different airports as well as over dinner, we discussed various international efforts and how other countries have addressed these issues in the past. Having studied housing, I have a long-standing interest in slum upgrading; I have been part of teams involved in the design of the Jawaharlal Nehru National Urban Renewal Mission and Rajiv Awaas Yojana (RAY) which aimed at citywide infrastructure and integrating slums into the city fabric.

In our discussions, it came up that even during the RAY programme, Odisha had taken the lead from the beginning and had drafted an act for housing and a slum-free city policy.

Following this, we were in regular contact trying to think about the next challenges that the Jaga mission could address. Thereafter, on a subsequent visit to Sri Lanka to attend the International Water Association conference, we got to discuss a framework for trying to move the Jaga mission from providing individual benefits of tenure security and improved housing to community benefits and city-level benefits (as shown in the diagram below). The idea is to provide community infrastructure and public service benefits by involving citizens in basic services creating civic and social infrastructure so that the level of services in slums is brought close to what the rest of the city enjoys, thereby making them a more integral part of the city. These initial discussions contributed to the slum upgrading programme under the Jaga mission.

– **Shubhagato Dasgupta**

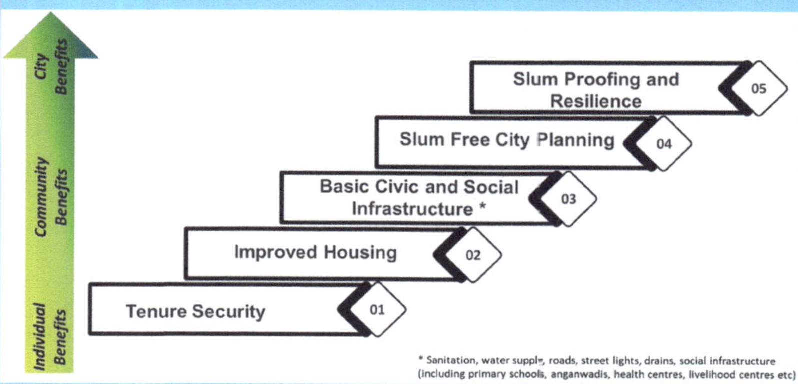

9.4 CREATING AN ENABLING ENVIRONMENT

The Department of Housing and Urban Development (HUDD) entered a memorandum of understanding with Scaling City Institutions for India to provide technical support in the phase 2 programme. A standard operating procedure (SOP) has been prepared to guide urban local bodies (ULBs) and SDAs in the process of delisting of slums through the creation of necessary infrastructure and bringing the slums at par with the rest of the city.

This SOP seeks to identify the key primary infrastructure gaps so that those gaps can be filled, and to foster community participation by involving SDAs in the processes of planning and development. It aims to decentralise the decision-making processes by empowering the ward officers. The SOP for participatory slum upgradation and delisting was launched on 28 September 2020. It laid down the procedure and benchmarks for undertaking participatory slum infrastructure upgradation, and the established procedure and documentation for delisting a slum to integrate it with the rest of the urban area.

To strengthen programme implementation through the infusion of financial resources, the Odisha government made an amendment in the Odisha Municipal Act, 1950, and Odisha Municipal Corporation Act, 2003, to earmark 25% of the municipal budget under capex for infrastructure development in the slum. The ULBs are required to consider this amendment while preparing their budget and mandatorily spend 25% of it in slum areas to provide basic services and infrastructure and on other ongoing schemes related to slums. The ULBs are also responsible for monitoring the spending on a quarterly basis.

9.5 BENCHMARKING BASIC INFRASTRUCTURE AMENITIES AND SERVICES TO ACHIEVE LIVEABLE HABITAT STATUS

To transform slums into liveable habitats, upgradation and delisting were taken up. Basic infrastructure and amenities – namely, tap water supply at the HH level, access to toilet facilities, paved roads, constructed storm water drainage, streetlights, in-house electricity connection, community centres called parichayas and creation of open space – have been taken up. To identify the gap in availability of the basic infrastructure and amenities in a slum, the following parameters are considered:

- **Water supply:** The presence of piped water supply infrastructure at the street level and of house connections in the slum is assessed to understand the gap in access to piped water supply at the individual HH level. Following this, the gap is assessed in terms of percentage of HHs on a street with water supply pipeline. This percentage is then used to formulate a detailed project report (DPR) for undertaking a water supply project in the slum. The Public Health Engineering Organisation/Water Corporation of Odisha is responsible for implementing the water supply project.

  % of HHs in the street with water supply pipeline = (*No. of HHs in the street having access to water supply pipeline/Total no. of HHs*) × 100

- **Paved roads:** For assessing the condition of the roads inside the slum, data on certain parameters – such as dwelling units with access to public road/street within the slum and the length of inner road/streets (in km) – are collected and then the percentage of total existing paved road is calculated to identify the gap. The DPRs prepared for roads include the use of paver blocks to enhance subsoil percolation and groundwater recharge. Road projects are undertaken by the ULBs.

  % of total constructed road length = (*Total paved road length*/ *Total road length*) × 100

- **Constructed storm water drainage:** To understand the gap in this infrastructure, the coverage of constructed drains is deduced by ascertaining the number of dwelling units connected to these drains. Construction of covered drains is encouraged in the slums because blocked drains and stagnant water lead to health hazards and groundwater contamination. Construction of drains is performed by the ULBs in collaboration with the SDAs.

  % Coverage of constructed drains = (*No. of HHs connected/discharging to constructed drains/Total no. of households*) × 100

- **Streetlights:** The projects on streetlights are formulated based on the gap analysed in terms of the number of functional streetlights and the street length covered by the functional streetlights. Light-emitting diode (LED) bulbs are used for the purpose of street lighting, to enhance energy conservation, reduce the carbon footprint and conserve fuel.

  % of street length with functional streetlights = (*Length of street with functional streetlight/Total internal street length*) × 100

- **Access to toilet facility:** To understand the gap and formulate projects for ensuring access to toilet facilities, several parameters are studied: dwelling units with a septic tank/pit/sewer system and functional individual household latrines (IHHLs), number of HHs having access to toilets (IHHL + CT/PT (community toilet/public toilet)), number of community toilets and number of seats available in community toilets. These parameters are subsequently used to calculate the deficit as percentage of HHs having access to a safe, functional toilet. DPRs prepared for the identified infrastructure gap under this vertical focus on climate-sensitive design, such as construction of a toilet with a septic tank and so on, to reduce groundwater contamination, water body contamination and health hazards, resulting in a positive environmental impact.

  % of HHs having access to safe functional toilet = [(*No. of HHs with IHHL for exclusive use* + *no. of HHs having access to CT/PT*)/*Total number of households in the slum*] × 100

- **In-house electricity:** Data on the number of dwelling units with individual authorised electricity connection are collected through focus group discussions to calculate the requirement for in-house electricity connections. In-house electricity is supplied by the electricity distribution company to the HH based on submission of application for connection. In-house electricity encourages energy conservation, reduction in carbon footprint and fuel conservation results in positive environmental impact.

  % of HHs with individual authorised electricity connection = (*No. of HHs with authorised electricity connections/Total no. of HHs in the slum*) × 100

- **Parichaya:** Parichayas are the multipurpose community centres accessible to all slum dwellers. There shall be at least one community centre (parichaya) within the slum or in proximity with unhindered access.
- **Open space development:** Every slum should have at least one open space development within the slum or in proximity, accessible to the slum dwellers.
- **Children play areas/zones:** There should be at least one children's play area within the slum or in proximity, accessible to the slum dwellers.

9.6 PROJECT IMPLEMENTATION

The way forward envisaged for slum upgradation and delisting is:

(a) Address the primary inequalities prevailing in urban areas to integrate the informal settlements into the mainstream city fabric.
(b) Create conditions for reducing the growth of slums in urban areas in the future, by addressing the core issues of access to land, services, housing and other necessities of urban life for the poor.
(c) Move up the ladder of building resilience in urban areas and slum-proofing the city by progressing from the HH level to the community level to the city level interventions, thereby ensuring increased urban resilience.

To achieve its vision of inclusive, egalitarian and liveable urban habitats, the Government of Odisha aims to ensure the participation of slum residents in urban planning, development and decentralisation of decision-making processes (by activating ward officers) through the SOP for slum upgradation and delisting. The steps have been detailed below:

(i) Identification and listing of slums
The ULBs identify all the slums within their jurisdiction and prepared a list of slums as mandated by Urban Area Slum Redevelopment and Rehabilitation Committee (UASRRC).
(ii) Undertake participatory infrastructure needs assessment (PINA) at the slum level for primary infrastructure facilities
Once the slums on tenable land are identified, slum-level PINA is to be undertaken to reveal the status of access to the primary infrastructure,

specifically (i) water supply, (ii) constructed roads, (iii) constructed storm water drainage, (iv) streetlights, (v) access to toilet facilities and (vi) in-house electricity. This survey forms the baseline for preparing the infrastructure gap assessment profile (i-GAP) at the next level.

(iii) Prepare i-GAP to assess the infrastructure gap at the slum level and plan for slum upgradation projects
An i-GAP is developed as an immediate next step, with information collected through PINA. This is prepared by the ULB in close coordination with the concerned ward officer. This tool is useful to objectively assess the vulnerability of the slums in terms of the availability of primary infrastructure and to develop an informed, multipronged and phasing strategy for delisting. If any slum has adequate infrastructure, it is to be directly taken up for delisting; otherwise, slum upgradation projects addressing the gaps are to be formulated.

(iv) Execution and completion of slum upgradation projects
The slum upgradation projects are executed by the concerned SDA under the supervision of the ward officer on behalf of the ULB. The concerned SDA is issued with the work orders for implementing the works as implementing partners (IPs) for such slum upgradation work.

(v) Delisting of slum
On completion of the projects, a resolution is passed by the SDA/RWA stating that the slum is eligible for delisting. The proposal is then submitted to the ULB by the SDA. It is verified by the ULB and passed on to the UASRRC for delisting. The delisting proposal is forwarded to the Delisting Scrutiny Sub-Committee (DSC) by the UASRRC to validate the slum upgradation work through field verification. Upon validation of the process of delisting, the UASRRC issues a public notice inviting objections within a period of 15 days to the change of status of the slum. After considering the objections, if any, UASRRC approves the delisting proposal and directs the concerned ULB to issue an order for delisting the slum. The concerned SDA subsequently passes a resolution converting the SDA to RWA, with an intimation to the ULB through the concerned ward officer (Figure 9.1).

9.7 INSTITUTIONAL ARRANGEMENTS

The slum upgradation and delisting programme has the involvement of stakeholders at all levels, from state to city, ward and slum, supported by Jaga coordinators, local non-governmental organisations (NGOs) and community-based organisations (CBOs).

9.7.1 At the state level

At the state level the HUDD is the nodal agency for implementing the provisions of the Land Rights Act (LRA) as well as the Jaga mission. The Chief Secretary of the state is the chairperson of the governing body. The Principal Secretary HUDD is the chairperson of the executive committee, supported by director, municipal administration, as the mission director.

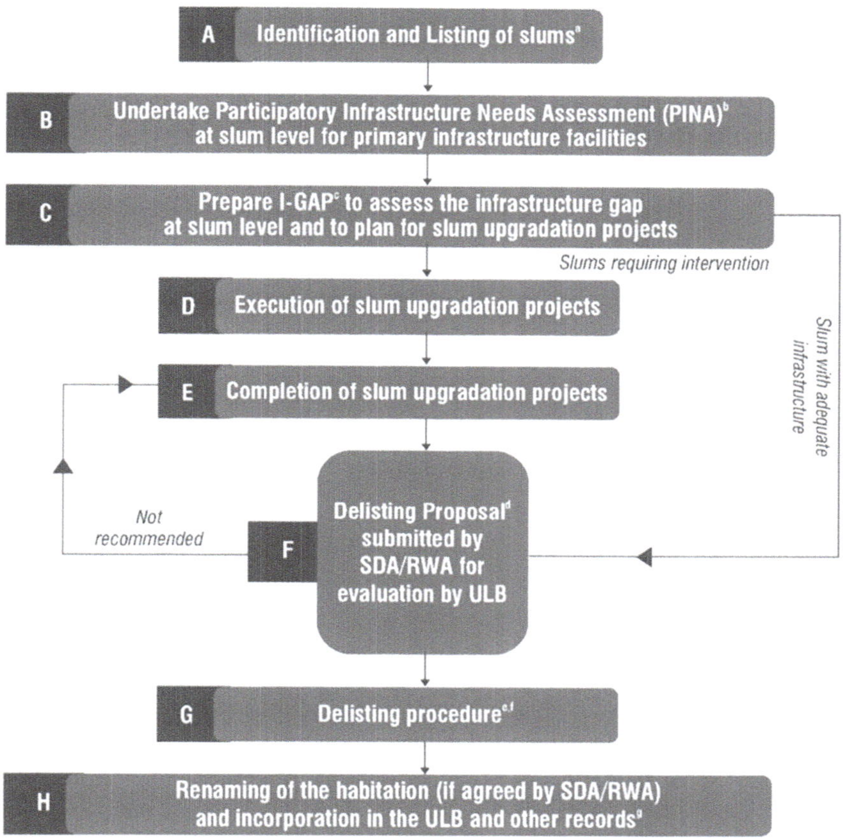

Figure 9.1 Process for slum upgradation and delisting. *Source*: SOP for slum upgradation and delisting.

9.7.2 At the ULB level

The *UASRRC* in each ULB is headed by the District Collector and constituted by the HUDD. The UASRRC is responsible for formulating the Scrutiny Sub-Committee and preparing the list of slum dwellers eligible for issuance of LRCs. The UASRRC typically consists of the following members:

- District Collector (Head)
- An elected representative of the ULB who is nominated by the respective municipal council
- Superintendent of Police (or a representative not below the rank of Deputy Superintendent of Police (DSP) or sub-divisional police officer (SDPO))
- Tehsildar (of the corresponding ULB)
- Executive Officer (EO) of the ULB

- One urban planner (who is nominated by the District Collector)
- One CBO/slum dwellers' representative (who is nominated by the District Collector)
- One NGO representative (who is nominated by the District Collector)

The UASRRC constitutes a *DSC* to validate the slum delisting recommendations received from the SDA/RWA through the ULB. The DSC consists of the following members:

- One senior officer to be nominated by the District Collector – chairperson
- President and Secretary of the SDA/RWA from the concerned slum – member
- One civil society representative nominated by the District Collector for the entire ULB – member
- One educationist nominated by the District Collector for the entire ULB – member
- Executive Officer (EO)/Commissioner/any senior officer of the concerned ULB – member convener
- Other invited members as the chairperson may deem fit

9.7.3 At the ward level

The concerned ward officer is the designated nodal officer for the process of slum upgradation and delisting in his/her jurisdiction.

9.7.4 At the community level

The *SDA* also plays an important role in the second phase of Jaga mission. The concerned SDA/RWA is the IP for these slum upgradation works as will be discussed in detail in Chapter 10. The SDA/RWA members facilitate community participation and support the ULB officials to gain insights into the infrastructure need and gap assessment. SDA is also responsible for preparing proposal for delisting the slum.

9.8 BUDGET AND FINANCING

The HUDD is implementing the programme by utilising the existing human resources at the state, district and ULB levels. As their establishment cost is already provided and met from the annual state budget, no additional budget provision is required. The Jaga mission is unique in its use of convergence as a tool in implementation, and hence the budgeting process. The various public works components are funded under various government schemes, converging and collaborating to cover all parts of the project. For example, when Jaga upgrades slums, it does not create a separate component for drinking water supply within its project parameters. Rather, it simply converges with Mission Buxi Jagabandhu Assured Water Supply to Habitations (BASUDHA), which has a dedicated fund. Similarly, LED street lighting is performed through the state LED programme under implementation at the pan state level. Another very important programme converged with for other infra works is MUKTA (erstwhile urban wage employment initiative (UWEI)) under which community-level infrastructure works can be undertaken in partnership with community-based organisations such as Mission Shakti, women's self-help groups and their area-/city-level federations and SDAs. SDAs as IPs are entrusted with works up to Rs 200 000. Works beyond this threshold are implemented directly by the ULBs, but engaging SDAs as supervision partners for which 7.5% of the project cost is paid as supervision charges to the SDAs.

9.9 LINKAGES WITH ONGOING PROGRAMMES/MISSIONS

As discussed above, the slum upgradation programme under the Jaga mission has been converged with various central and state government schemes. It is complemented by the state schemes BASUDHA, that seeks to provide piped drinking water supply to all HHs, and Mission Unnati, that manages the convergence of all urban schemes. Jaga has been converged with the MUKTA yojana – the erstwhile UWEI – to provide livelihood opportunity to slum dwellers by engaging them in the construction of the infrastructures for slum upgrading. Pradhan Mantri Awaas Yojana (Urban) – also known as Odisha Urban Housing Mission – and the Jaga mission work in close collaboration with ensure that land rights and housing are made available to slum dwellers. Central schemes such as Swachh Bharat Mission and Atal Mission for Rejuvenation and Urban Transformation have also been converged to provide water and sanitation in slums.

9.10 ON THE MARCH TO COMPLETION

Phase 2 of the Jaga mission has shown excellent results. In March 2021, it was decided to roll out the slum upgradation and delisting process for 500 slums. However, in just 6 months, 550 slums had been upgraded and delisted across 28 ULBs. Currently about 1654 slums have been upgraded and delisted across 115 ULBs. Of these, 28 have been declared slum-free (see Table 9.1), meaning every area of those cities is at par when it comes to amenities and civic services.

The goal of slum-free and slum-proofed cities is well underway with the concurrent implementation of both phases of the Jaga mission. As that target nears, it will become clear if more evolution is needed under this banner to maintain and run Odisha's cities with a high level of inclusivity, integration and resilience.

One of the learnings from the implementation of phase 2 of Jaga is that though the state has a few schemes to provide basic infrastructure, amenities and services, and has earmarked sufficient budget provisions to achieve universal coverage, initiatives such as the Jaga mission are still vital to ensure provision of basic infra and services. Phase 2 provides empirical evidence to that effect. The state is implementing the BASUDHA mission to provide universal coverage of piped water supply connection with an investment of about Rs 4000 crores; there is a smart LED street lighting programme with an investment of Rs 750 crores. Although every HH in a slum should have got piped water connection under BASUDHA and every street of the slum should have got smart LED street lighting, PINA and i-GAP analysis under the Jaga mission revealed that these programmes could not adequately penetrate the slums – not because the state did not want the slums to be covered but

Table 9.1 List of pilot ULBs undertaking slum upgradation and delisting until March 2021.

Sl. No.	Name of City	Sl. No.	Name of City
1	Angul	2	Kotpad
3	Boudhgarh	4	Koraput
5	Sundargarh	6	Konark
7	Khurdha	8	Gopalpur
9	Chikiti	10	Behrampur
11	Polasara	12	Bhubaneswar
13	Ganjam	14	Cuttack
15	Jatni	16	Paradeep
17	Digapahandi	18	Aska
19	Hinjilikut	20	Kavisurya Nagar
21	Sunabeda	22	Kodala
23	Dhenkanal	24	Chatrapur
25	Bellaguntha	26	Buguda
27	Soroda	28	Rourkela

Source: Government of Odisha.

because of the systemic reluctance and deeply ingrained bias against the poor in the bureaucratic system. Without the focused and dedicated approach under the Jaga mission, the dream of universal coverage of basic infrastructure for the poor would not have been possible.

9.11 EXTENDING LAND RIGHTS TO SLUM DWELLERS SETTLED ON NON-GOVERNMENTAL LANDS: A CRITICAL BUT CHALLENGING TASK

Developing processes for transferring land rights to slum dwellers where the land itself does not belong to the municipality is one of the most critical and difficult challenges that policy-makers face in slum land rights and upgrading programmes. This challenge became evident in Odisha, too, after the execution of the first few rounds of the land rights programme.

The Jaga mission team set about understanding the relevant issues bottom up through detailed mapping of the slum settlements and of the land ownership. Before this detailed study of land issues occurred, the basic understanding was that people with LEC will be given a new dwelling unit and will not be settled at the same location where their slums were. This however was found to be one of the important weaknesses of programmes that address slum housing, as relocation of slum dwellers causes major disruption to their livelihood and work–life balance, often pushing them further into the poverty trap than raising them out of poverty.

To minimise relocation of slum dwellers who are not settled on government land and could not initially be given LRCs, the slums in Odisha were categorised into a set of categories (as shown below); various planning instruments and land exchange instruments are in the process of being deployed to resolve the

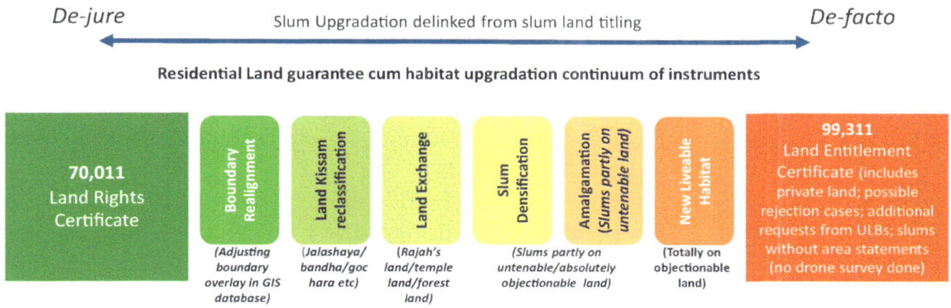

Figure 9.2 Categories of slums. *Source*: Authors.

issue in each slum. This is a difficult but essential exercise and needs to happen at the slum level in consultation with both the ULB and district office including other land-owning departments. Some examples of where these exercises have been successful are discussed in this chapter. It is important to note that Odisha is possibly the first example of a state that is systematically redefining land entitlement based on scientific and detailed processes. Many states have a lot to learn from this example, be it in the areas of land exchange, re-densification or boundary realignment or even reclassification.

9.12 STRATEGIES TO CONVERT LECS TO LRCS

During the urban slum household area (USHA) survey, half of the slums were found to be on lands that were untenable, with the residents living in environmentally hazardous locations or under significant threat of eviction. These slums had come up on environmentally hazardous land such as those close to garbage dumps; land owned by the railways, forest or defence departments; lands reserved in the past by the Revenue Department for water bodies, flood plains or grazing land and lands owned by local erstwhile kings of princely states or temple committees. To address untenability of land, re-densification of existing tenable slums and relocation to a new liveable habitat are being explored. However, to avoid large-scale relocation, the state emphasises alternatives such as land exchange with private entities and/or conversion of land classification to make the land parcels leasable. Only after exhausting all such options are the ULBs allowed to choose the relocation option, which is primarily meant for slums located on the strictly objectionable category of land.

9.12.1 New liveable habitats

For the people at the margins of the cities occupying environmentally hazardous land or those living under significant threat of eviction, the government has developed an SOP for creating new liveable habitats. The principles guiding the present process of relocation to a new habitat are:

(a) Exhaustion of all alternatives for in-situ settlement, including re-densification of the existing tenable slums, before exploring relocation to a new liveable habitat as the option.

(b) *En masse* relocation of the slum HHs which lie wholly in the 'objectionable' category of untenable land.
(c) Relocation of the slum HHs to be subjected to their prior consent obtained through a resolution signed by each individual HH.
(d) Leveraging available housing schemes to encourage beneficiary-led housing at the new liveable habitat.
(e) Creation of a liveable habitat with all basic infrastructure, amenities and services in government land along with providing individual plots with land rights to every family and linking them with housing schemes.

The option of creation of a new habitat applies to all municipalities and NACs in Odisha, and relocation of slums to new habitats has been undertaken for slums on the 'strictly objectionable' category of untenable land. The 'objectionable' categories of untenable land include but are not limited to:

(a) Railway land.
(b) Defence land.
(c) Any other state government land that cannot be transferred inter-departmentally to be utilised for settlement of land rights to the dwellers and creation of basic civic infrastructure.
(d) Environmentally hazardous land that is not fit for human habitation and entails undue risk to the safety or health or life of the residents or habitations, such as those near waste dumping sites, low-lying areas, land prone to landslides, subsidence-prone areas and so on, as identified and declared by ULBs and ratified by UASRRC.
(e) Any other type of land as may be notified by the state government from time to time.

The new liveable habitats are planned as plotted mini-townships, with key primary infrastructure facilities – including individual houses with toilets connected to a community septic tank, well-designed and paved streets, piped water supply (in-house), street lighting, in-house electricity connection, storm water drainage and necessary social infrastructure such as parichayas, open space development and children's play areas. As the new habitats are created as green field projects on government lands, availability of sufficient land for these social infrastructures are ensured.

Definitions

(1) Non-objectionable category: It may include, but is not limited to, forest land, private land (mostly local Raja's land), temple land or any other land that may be converted to tenable by land exchange/negotiation or conversion.
(2) Objectionable: It may include, but not limited to, railway land, defence land, environmentally hazardous land, land near waste dumping sites and so on or any other land where existence of

human habitation entails undue risk to the safety or health or life of the inhabitants themselves or where habitation on such areas is considered by the committee not to be in public interest. Smaller settlements having less than 20 HHs and does not qualify as slum can also be included in this category.

Source: SOP for New Liveable Habitats.

The guiding design principles for the new habitats include:

(a) The plots in the new liveable habitat shall be earmarked for residential purposes only.
(b) The project scale will be commensurate with the number of HHs to be relocated.
(c) Open space and common social infrastructure such as parichayas shall be placed preferably at a central location to allow all the HHs to have easy access.
(d) Construction of common septic tanks catering to three to four individual houses shall be encouraged; a suitable area may be demarcated for this. Such demarcated spaces (above the common septic tanks) can be used as an open area for the public.
(e) Streets will have adequate width to allow easy access to emergency services, for example, ambulance, fire tender and so on.

The new SOPs for the establishment of new liveable habitats have been taken up as pilot projects in seven cities – Kesinga, Dhenkanal, Bolangir, Soro, Balasore, Gopalpur and Jaleswar – with active support from the stakeholders. Each of these cities is at a different stage of implementation, Kesinga being the most advanced where construction work has already begun. The expansion of the new liveable habitat segment of Jaga can be expected after fine-tuning of the SOPs based on the experience in these seven cities; this will be scaled up to 30 cities. Out of a total of about 1200 slums affected by various degrees of untenability, only about 200 or so are being considered for resettlement now. Work is in progress to reduce even this number in the future.

9.12.2 Forest land diversion

The Odisha government amended the Forest Rights Act, 2006 to divert forest land to create infrastructure for slum dwellers on land classified as forest land. As per the issued guidelines on 'diversion of forest land for construction of facilities at the urban slums situated over Forest land under the Scheduled Tribes and Other Traditional Forest Dwellers Act, 2006', in the first phase, the SOP extends to all slums on revenue forest land and all slums covered under the USHA survey. It is applicable to lands in various categories (Kissam), such as Gramya Jungle, Pathara Jungle, Chota Jungle, Jhati Jungle and so on, in municipalities and NACs in Odisha. An SOP was worked out jointly between the

Forest and Environment Department and the Urban Development Department, which was piloted in the Dhenkanal Municipality. Diversion of forest land has been performed for the development of the following infrastructure facilities within slum boundaries:

(a) schools
(b) dispensary or hospital
(c) anganwadis
(d) fair price shops
(e) electric and telecommunication lines
(f) tanks and other minor water bodies
(g) drinking water supply and water pipelines
(h) water or rainwater harvesting structures
(i) minor irrigation canals
(j) non-conventional source of energy
(k) skill upgradation or vocational training centres
(l) roads
(m) community centres/parichayas

All 235 slums identified to be on forest land are in various stages of transition.

9.12.3 Other strategies for in-situ slum upgradation

Subsequently, the government has identified another 59 slums across 28 ULBs that can be taken up for re-densification. Re-densification is the process of densifying a slum situated on a tenable land provided there is land available for settlement. For this process, written consent from concerned SDAs and concerned beneficiaries is to be obtained. Currently, re-densification is being piloted in the ULBs of Odogaon, Hindol and Ranpur where 15 HHs or less are residing on forest land. For scale-up of forest re-densification work, identified ULBs are presently finalising non-forest government vacant land within or adjacent to the slum.

To maximise in-situ settlement of land and minimise relocation, the HUDD has been exploring other strategies on a case-to-case basis, such as slum boundary readjustment, re-survey of un-surveyed HH, land exchange and negotiation. These cases are being piloted based on which SOPs will be prepared in future.

9.13 SLUM-PROOFING ODISHA: A BLUEPRINT FOR SUSTAINABLE URBAN DEVELOPMENT

In the pursuit of sustainable urban development, the issue of slums serves as a stark reminder of the challenges cities face worldwide. Slums not only signify a failure to provide adequate housing and basic services to a significant portion of the population but also act as breeding grounds for social inequalities, health hazards and economic disparities. Despite efforts to upgrade existing slums, the underlying causes of slum formation persist, necessitating proactive measures to prevent their recurrence. Slum-proofing urban areas in Odisha goes beyond merely making cities slum-free, it entails addressing the root causes of slum formation and ensuring the sustainable development of urban areas. Enter the concept of 'slum-proofing' cities – a strategic approach aimed at tackling the core issues of slum formation and fostering inclusive, resilient and sustainable urban growth.

9.13.1 Why slum-proofing cities in Odisha?

The urgency to slum-proofing cities in Odisha arises from the recognition that traditional approaches to urban development have often been reactive rather than proactive. Although slum upgrading initiatives have their merits, they tend to focus on ameliorating existing conditions rather than preventing their recurrence. By shifting the paradigm towards slum-proofing, cities can break the cycle of poverty and deprivation perpetuated in informal settlements. Moreover, slum-proofing aligns with broader sustainability agendas, including the United Nations sustainable development goals, particularly goal 11: Sustainable cities and communities.

Slum-proofing cities in Odisha are crucial for several reasons:

- Rapid urbanisation: Odisha is witnessing rapid urbanisation, with a significant influx of migrants from rural areas seeking better opportunities. This influx, coupled with natural population growth within cities, contributes to the formation of new slums if left unaddressed.
- Sustainable development: Slums are symptomatic of broader systemic challenges such as inadequate urban planning, infrastructure and access to basic services. Slum-proofing initiatives are essential for promoting sustainable urban development in Odisha, ensuring inclusivity, resilience and meeting the needs of all residents.
- Community empowerment: Slum-proofing efforts prioritise community participation in urban planning and decision-making processes. By empowering residents to voice their needs and priorities, cities in Odisha can foster ownership and accountability, leading to more effective and sustainable development outcomes.

9.13.2 How we are slum-proofing cities in Odisha?

To slum-proof cities in Odisha, a strategic and tailored approach is imperative, considering the state's unique context and challenges. Here are some specific actions taken to realise this goal:

(1) **Formation of State-Level Monitoring Committee (SLMC):** Establishing a committee at the state level to oversee and coordinate slum-proofing initiatives, ensuring alignment with broader development objectives.
(2) **Creation of state-level portal for registration of urban poor and migrants:** Developing an online platform to streamline the registration process for urban poor and migrants, facilitating access to social and affordable housing options.
(3) **Creation of City-Level Migration Committee (CLMC):** Setting up committees at the city level to address migration-related challenges and coordinate efforts to provide housing and support services to migrants.

(4) **Existing ward-level committees' support**: Leveraging existing ward-level committees to assist city-level migration committees in identifying and addressing the needs of migrants.
(5) **Enumeration of existing accommodation facilities and creation of land bank:** Conducting surveys to assess existing accommodation facilities and government-owned vacant buildings for potential conversion into housing units. Additionally, establishing a land bank to identify available land for future housing developments.
(6) **Capacity building and public awareness campaigns:** Conducting capacity-building programmes and awareness campaigns to educate stakeholders about migrant registration processes and available housing options.
(7) **Registration of migrants:** Establishing help desks and online registration facilities at municipality offices to streamline the registration process for migrants. Disseminating information through advertisements and pamphlets at key locations frequented by migrants.
Benefits of Registration
(a) Early allotment: Registering early ensures prompt allocation of housing, addressing immediate shelter needs.
(b) Access to anganwadis: Provides vital early childhood care and education, benefiting child development.
(c) Access to schools: Facilitates enrolment in schools, ensuring educational opportunities for children.
(d) Access to public distribution system (PDS) rations: Ensures access to subsidised food, addressing food insecurity and promoting well-being.
(8) **Assessment of housing and infrastructure demand:** Conducting comprehensive assessments to determine current housing and infrastructure deficiencies and projecting future demand to bridge the gap effectively.
(9) Provision for housing options:
- **Social housing:** Utilising existing shelter facilities or government buildings for temporary accommodation, retrofitting or redeveloping them to meet the needs of migrants.
- **Affordable housing units**: Developing affordable housing units with financial assistance from state and central housing schemes, ensuring allocation to eligible beneficiaries based on the Odisha Housing for All Policy for Urban Areas, 2022.

(10) **Approval and allotment:** ULBs and CLMCs approving and allocating housing options to individual beneficiaries, ensuring equitable distribution and adherence to established policies and guidelines.

By implementing these measures, Odisha aims to slum-proof its cities, ensuring that all residents have access to adequate housing and essential services, regardless of their socio-economic status or background. To implement the strategy of slum-proofing in urban areas, an SOP is being laid out for operationalisation in Odisha.

9.13.3 What is slum-proofing?

Slum-proofing refers to a holistic approach to urban development aimed at addressing the root causes of slum formation and preventing the emergence of new slums in cities. It involves implementing strategies to improve access to housing, services and infrastructure for marginalised communities, while also promoting community participation and resilience-building measures.

This includes:

- **Registration and housing**: Establishing procedures and documentation for the registration of migrants in need of social and affordable housing is essential to ensure that vulnerable populations have access to adequate housing options.
- **Addressing core issues:** Tackling the underlying factors contributing to slum formation requires a multi-faceted approach. This may involve land-use planning, infrastructure development and affordable housing programmes targeted at marginalised communities in Odisha.
- **Building urban resilience:** Slum-proofing initiatives should aim to build resilience at the individual, community and city levels in Odisha. This includes measures such as disaster preparedness, access to social safety nets and community-based development initiatives that empower residents to cope with shocks and stresses.

9.13.4 Outcomes and impact

Slum-proofing Odisha's cities holds the promise of transformative change, with potential outcomes including:

(1) **Social inclusion:** Promoting social cohesion and equity by ensuring that all urban residents have access to decent housing, basic services and opportunities for socio-economic advancement.
(2) **Economic development:** Unlocking the productive potential of marginalised communities, contributing to economic growth, job creation and poverty reduction.
(3) **Environmental sustainability:** Minimising resource consumption, mitigating the environmental impacts of urbanisation and enhancing resilience to climate change.
(4) **Resilience to shocks:** Strengthening infrastructure, social networks and institutional capacities to withstand various shocks and stresses, including natural disasters and socio-economic crises.

In conclusion, slum-proofing Odisha's cities is not only a moral imperative but also a strategic necessity for achieving sustainable urban development in the state. By addressing the root causes of slum formation and implementing proactive measures to prevent their emergence, cities can create environments where all residents can thrive. Through community participation, integrated planning and capacity building, Odisha can pave the way for a brighter and more equitable urban future.

Eviction

Source: HUDD, GoO.

Upgraded slum

Source: HUDD, GoO.

Upgraded slum

Source: HUDD, GoO.

Upgraded slum

Source: HUDD, GoO.

doi: 10.2166/9781789064926_0159

Chapter 10

Seeding social change by underwriting dignity and safety of sanitation workers: Garima scheme

ABSTRACT

This chapter on the Garima scheme provides details of how the government of Odisha has boldly recognised the existence of manual scavenging practices and decided to end it and to uplift the core sanitation workers engaged in that by ensuring safety, dignity and welfare. It is the first of its kind scheme launched by any state in India and is being followed by other states and has become the inspiration for the NAMASTE, a national scheme being implemented by Government of India.

Keywords: core sanitation worker, manual scavenging, stigmatisation, dignity, safe sanitation, social security

10.1 A PERSISTENT PRACTICE

Manual scavenging is arguably among the worst manifestations of a highly unequal social order characterised by exploitation, exclusion and untouchability. Despite being outlawed through various layers of legislation, the practice continues unabated across India today in various forms. Pronouncements of the Supreme Court and High Courts making the practice illegal have not succeeded in putting a stop to this heinous practice. Several factors such as oppressive, societal ostracism, systemic apathy, ignorance about the hazards it involves, media insensitivity and so on. have contributed to continuation of this reprehensible practice.

What is manual scavenging?

Manual scavenging is the act of handling, clearing, cleaning or disposing of human excreta from toilets, sewers or sceptic tanks. People who carry out these tasks usually do not have access to any manner of specialised equipment or protective gear. This means they either must handle excreta with their bare hands or enter drains and sceptic tank bare-bodied. Given the location of their work and the stigma attached, they rarely have access to sufficient water to clean themselves after finishing their task.

Manual scavenging is prohibited in India but persists insidiously.

As a task without dignity, manual scavenging has for millennia been part of India's sanitation scenario. Not long ago these activities were largely restricted to the evacuation of night soil. But it has taken on a hazardous and at times lethal edge with the advent of closed spaces in sanitation, such as septic tanks and underground sewerage systems. Unfortunately, successive governments have done little to eradicate the practice. This is not because technology on this front has not progressed enough to provide an alternative, but that adequate priority has not been given to investments to mechanise the cleaning process. If the core sanitation worker (CSW) stops performing these tasks our cities start stinking and will risk becoming massive health hazards in a span of just a day. Even then their importance and their criticality in the operation and maintenance of the city's sanitation system are not given their due

In India's social hierarchy, manual scavenging is typically performed by people belonging to Dalit communities. Even among Dalits, who have historically been subjected to the atrocities of untouchability, the sub-communities that carry out this critical task are treated with ignominy as the lowest within their hierarchy.

Manual scavenging is continuing despite being outlawed by the Constitution and laws, as well as decisions from the Supreme Court and High Courts, because our society hasn't yet developed a collective conscience against it.

10.1.1 It wasn't me

The social system of assigning the task to a particular community has only helped its continuation by insulating everyone else from it and by creating an oppressive web of circumstances that creates a vicious cycle and prevents escape, making it no better than slavery. The system acts as a poverty trap exploiting generation after generation.

Those who use manual scavengers including government, its agencies and private organisations and individuals seek a certain amount of indemnity by not engaging manual scavengers themselves, but by handing over the task to informal and unregulated intermediaries who in turn get the work done by the manual scavengers. This not only seemingly gives them legal cover, but also helps them escape the inherent immorality of it. This layer of protection allows individuals and institutions to claim innocence and escape liability whenever questions are raised, leading to a scenario where, officially, manual scavenging becomes invisible and non-existent on records. Officials have no hesitations in declaring to courts or other statutory bodies that manual scavenging does not exist in the areas under their administration. And they can do this because, technically, they have outsourced the work to the contractors and formal and informal agencies.

10.1.2 Dehumanising occupation

Those trapped in this system of sociological disorder often find themselves strongly stigmatised and with little in terms of education or capital to engage themselves in other occupations. The indignity of the work and the ostracisation apart, they also face acute hazards especially while working in confined spaces.

Cleaning and de-blocking of sceptic tanks and sewer systems pose the threat of exposure to harmful gases that could harm them and even kill them. And even in the cleaning of open drains, they face the risk of injuries to the head and limbs, dizziness and burning eyes. They also tend to develop acute skin infections and diseases since they must immerse themselves in fecal matter without any layer of protection. Lung diseases and other related ailments too are common resulting in reducing the average lifespans drastically.

In addition, they are subjected to psychological distress too, with workers reporting that their own children would not come closer to them due to the bad smell emanating from them that even soap could not wash off.

10.1.3 Passive collusion

While various pieces of legislation and rules have mandated that the hazardous cleaning of septic tanks and sewers should be carried out only with the required machinery following the prescribed procedure and using the personal protective gears this has hardly translated into reality as evident from frequent deaths reported in septic tank and sewer cleaning operations. Penal provisions against the engagement of persons as manual scavengers have failed to yield the intended outcome and eradicate this problem. This is a crime that remains invisible until a worker dies in the course of the work that also fades away after some time and goes unpunished except for some financial compensation.

Urban local bodies (ULBs) or other institutions are often not equipped with machines, protective gear and safety devices to clean the sewer system and septic tank. This means the workers must enter underground sewerage lines through manholes or confined space, putting themselves at risk of death by a fatal accident, asphyxiation because of lethal gases and harmful chemicals.

The scene of manual scavenging in Indian cities is not an uncommon phenomenon. However, governmental/private institutions and citizens collectively deny its existence. Despite several rounds of surveys taken up at the government level to identify manual scavengers and efforts to eliminate this inhumane practice, governmental agencies continue to declare that manual scavenging does not exist. As a result, with the existence of the problem not being acknowledged, it has become almost impossible to address it. The governments and the society both are part of a passive collusion system that keeps this relic of the past alive today.

10.1.4 Need for the scheme

Because of prolonged exposure to faecal matter while performing sanitation jobs like desludging of sanitation systems, entry into septic tanks/the sewer network for decongestion or removal of blockage or maintenance, CSWs come in direct contact with hazardous biological and chemical components present in the sludge and the toxic gases emanating therefrom causing health hazards, injuries and even death. Odisha was bold to acknowledge in public that the system of manual scavenging does exist in practice. It has been loud and open to discussing manual scavenging both within and outside on many platforms/forums in the recent few years upon acknowledging its importance. This boldness paved the way towards finding solutions for the problem.

In the beginning of 2019, Housing and Urban Development Department (HUDD) had a discussion with Urban Management Centre (UMC) on the very issue and idea to devise the scheme followed by organising a consultation with key organisations and activists: UMC, National Convenor of Safai Karamchari Andolan, Jan Saahas and WaterAid India. In addition, SKA Consultancy was consulted to understand the issues in depth. Sanitation workers' demands for improving socio-economic conditions were kept at the centre of the discussions, as was the need to ensure that core sanitation work is made a safe and dignified activity.

In these consultations, a series of issues were highlighted. It was pointed out that blockages in sewers many a time require manual cleaning, exposing workers to a variety of hazards including harmful toxic gases. Often the entire body of sanitation workers is exposed to faecal matter. They bathe multiple times every day out of compulsion, and yet the stench remains. They have limited access to hot water and have to use cold water even in winters for a bath after entering a sewer. Sanitation workers do not have access to changing rooms and water, sanitation and hygiene (WASH) facilities at work. The safety of sanitation workers while performing the specific task assigned to them and the climate are not taken into consideration while designing their personal protective equipment (PPE) or protective uniforms. The limited available protective gear is mainly intended for men. Women sanitation workers find it challenging to use this gear. Skin disease is common among such sanitation workers. Their children do not like to come near them. They face residential vulnerability. They live on the margins and often face social ostracisation. There continues to be aversion amongst community members in their children marrying into a family of sanitation workers.

Based on these consultations, the team working on drafting the scheme created a long wish list to address the safety, dignity and socio-economic issues confronted by sanitation workers. In this long aspirational list, problems and concerns voiced by the sanitation workers and their associations were prioritised. These eventually fed into the service level and social security benefits listed in the scheme.

The state government has committed to undo the generational injustice caused to this section of the society and made it clear that money should not stand in the way. The political and administrative will for improving the working and socio-economic conditions of these CSWs have been clearly demonstrated by accepting the scheme designed based on the recommendations emerged out of the consultation.

10.1.5 Uniqueness of the scheme

The Government of Odisha introduced the first of its kind scheme launched by any state in India for the safety, dignity and welfare of CSWs in 2020. The term 'Garima' denotes 'dignity' in the local language Odia, and through various provisions, the scheme aims at ensuring safe working conditions for sanitation workers and making this work a dignified profession. It also seeks to improve the socio-economic conditions of sanitation workers and their families.

The scheme provides an enabling environment to regulate core sanitation services. At the same time, it provides a comprehensive package that ensures service-level benefits, social security and financial benefits to the CSWs and their families.

10.1.6 Coverage

This scheme is being implemented in all 115 ULBs of Odisha. The Government of Odisha has set up a corpus fund with an initial amount of Rs 500 million dedicated to enhancing safety and dignity of about 20 000 (based on rough estimation) CSWs and their families benefitting a population of about 100 000. However, the government has committed to provide the required funds based on actual demand while implementing.

The scheme makes a nomenclature differentiation between 'Sanitation Work' and 'Core Sanitation Work'. Sanitation work includes activities such as street sweeping, garbage collection, cleaning of drains, toilets, sewers, septic tanks and so on. The Garima scheme covers only CSWs who deal with human faecal waste. Under Garima, CSWs include: (i) desludging of onsite sanitation systems, (ii) maintenance of sewerage network, (iii) cleaning of drains, (iv) operation and maintenance of sewage treatment plants (STPs) and faecal sludge treatment plants (FSTPs) and (v) cleaning, operation and maintenance of public and community toilets which involves dealing with human faecal matter in varying degrees. With the introduction of the scheme, CSWs are now included in scheduled employment and classified as 'Grade-I Highly Skilled' and 'Grade-II Skilled' workers considering the hazardous nature of this work and the untenable conditions in which they work every day.

With the Garima scheme, Odisha has become the first state to come forward and acknowledge the existence of unsafe cleaning of septic tanks and sewers while laying a clear path for how to resolve this problem, for which both the government and society at large share responsibility.

10.2 CREATING AN ENABLING ENVIRONMENT

Over the last six years, the state government has strengthened the legal framework to create an enabling environment to safely manage liquid waste and to safeguard workers engaged in sewer and septic tank cleaning. This has occurred through the introduction of the Odisha Urban Sanitation Policy and Odisha Urban Sanitation Strategy, 2017.

However, non-recognition or lack of regulation of informal service providers means that issues related to health, environment and exploitation of the workers remain unaddressed. Further, sanitation workers are forced to work in unsafe environments and are unable to raise their voice against blatant violations of the provisions of laws that are designed to protect their rights and interests.

With the introduction of the Garima scheme, the state has taken a series of measures to ensure that sewer or septic tank deaths no longer occur within the state. Individual responsibility has been fixed as per the law, and in case of a death, those found violating the law shall be prosecuted for criminal liability. To address safety, an Emergency Response Sanitation Unit (ERSU) has already been set up in each ULB. Initially, ULBs were entrusted with ERSU responsibilities, but because parastatal organisations like Public Health Engineering Organization (PHEO), Water Corporation of Odisha (WATCO) and Orissa Water Supply & Sewerage Board (OWSSB) play a significant role in providing important core sanitation services in the cities, the HUDD has decided that these organisations should take up the ERSUs' larger operational responsibilities where they have the presence of an officer in and above the rank of assistant executive engineer (AEE). With the above arrangement, the state currently has 77 ERSUs led by ULBs, six led by WATCO, 30 led by PHEO and two led by OWSSB.

Each city has an ERSU in place and the Responsible Sanitation Authority (RSA) has been designated. ERSU's primary duty is to make sure that sanitation workers are safe while performing their jobs. If there is a cleaning and maintenance issue related to septic tanks and sewers, citizens/service seekers call at 14420, a universal toll-free helpline number used by ERSU in all cities. The telephone operator handling the helpline number registers the request in the prescribed format and directs it to the designated duty supervisor from the ULB.

If the request is for mechanical sewer cleaning/desludging, the duty supervisor will ensure that the required equipment, including safety devices, is available at the work site and assign a registered Private Sanitation Service Organisation (PSSO) with a 'Command Certificate' to do the cleaning activity with the help of sewer entry professionals (SEPs) – only those who are trained and certified – using appropriate machines and wearing proper PPE. Post completion of the activity, PSSO will submit a 'Completion Report' to the duty supervisor.

In case mechanical cleaning is not possible, the duty supervisor will visit the site and prepare a Hazard Assessment Report. The duty supervisor will take permission from the RSA by submitting this report. RSA will analyse the report and give permission in writing for manual cleaning in all such cases where human entry is unavoidable, duly recording the reason for the same. After getting permission from RSA, the duty supervisor will inform the ambulance, police and fire service to ensure the safety of sewer entry professionals (SEPs) during and after the cleaning activity. The duty supervisor will take consent from the SEPs through a 'Self-Declaration Format', after making them aware of all the risks and hazards involved.

The duty supervisor must ensure that proper SoP is followed, all PPEs are worn, and safety devices are in place to avoid accidents on the site. He/she must

Figure 10.1 Process flow as prescribed under ERSU. *Source*: Emergency Response Sanitation Unit.

be present at the site throughout the period the cleaning/maintenance process is carried out. Post completion of the activity, the duty supervisor will submit a Completion Report to RSA (Figure 10.1).

10.3 LAYING THE GROUNDWORK

The scheme envisages a state-wide survey to identify sanitation workers engaged in core sanitation work and regulating core sanitation service organisations including those in the private and informal sector and introduction of a special wage category.

A pilot survey was carried out in six large cities, followed by a scale-up survey in the remaining 109 cities of the state to identify the CSWs. Often these workers are engaged part-time in sanitation work. For a worker to be identified as a CSW, they should have undertaken tasks listed as core sanitation work in the last 50 days. After being identified, these workers are validated; the validated CSWs are given a Garima ID. Each worker's unique ID will serve as their one-stop information point for all benefits, including social security benefits and service-level benefits. The survey is crucial for identifying the genuine workers performing core sanitation work who are to be covered under the scheme to ensure that various safeguards and benefits provided under Garima reach the right beneficiaries and not those who may be sanitation workers but are not performing core sanitation work, that is handling human faecal matter, as defined under the scheme.

To provide core sanitation services, individuals and PSSOs need to be registered with the ULB. These services will be rendered only by registered CSWs under the supervision of the ULB/PSSO to ensure safety in service delivery. Provision of protective equipment and safety devices and appropriate cleaning equipment to CSWs is the responsibility of the ULB/PSSO.

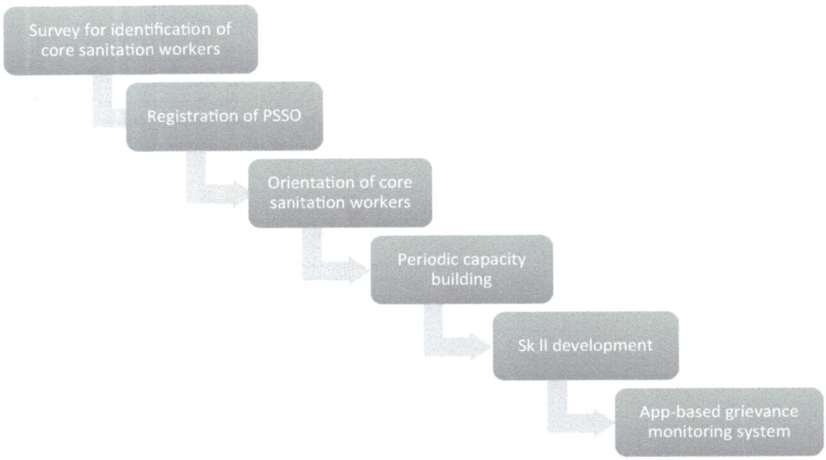

Figure 10.2 Process to ensure efficient service delivery to CSWs. *Source*: GARIMA scheme document.

After CSWs are identified and registered and PSSOs are empanelled, orientation, periodic capacity building and skill development of the workers are initiated to ensure efficient service delivery.

A helpline number has been set up for the use of both citizens and CSWs at each city level. A citizen can use this portal for seeking the service and CSWs can use it for registering a complaint if they are not getting the assured wages, protective uniform or safety devices.

The government is exploring mechanisms to provide easy access to government benefits available to CSWs. This would also be crucial for linking CSWs to ongoing schemes and benefits announced by the central and state government across various departments (Figure 10.2).

The scheme aims to ensure safety in Sanitation service delivery. Under the scheme, sanitation services in confined space will be closely monitored. More attention has been placed on ensuring 100% mechanised cleaning, and provisions for PPE, safety devices and appropriate cleaning devices are becoming non-negotiable.

10.3.1 Service benefits

CSWs will have access to a set of service benefits including health and life insurance, risk and hardship allowance (Figure 10.3).

10.3.2 Social security benefits

Other benefits include priority in access to educational institutions for the children of the CSWs, 90% grant in housing assistance, 90% grant for buying a two-wheeler, mandatory provision of PPE and safety devices, ex-gratia in case of accident and injury, and access to Life Insurance, Employees' Provident Fund (EPF), Employees' State Insurance (ESI) and family pension.

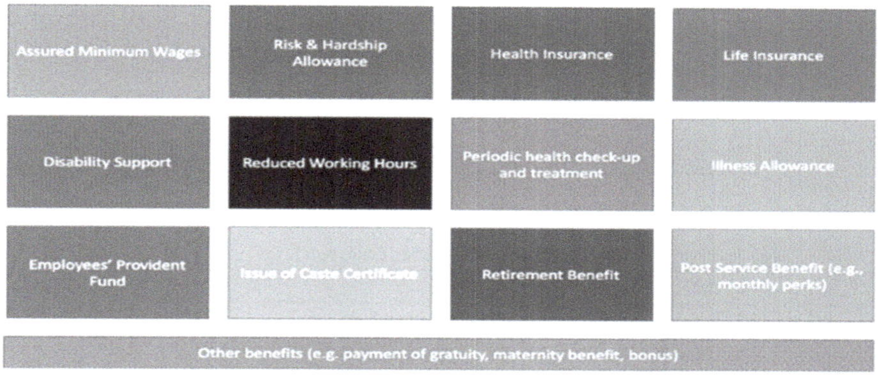

Figure 10.3 Assured service benefits for CSWs. *Source*: Garima scheme guideline.

Some benefits like hikes in wages, safety devices, free mobile phones and subsidy for two-wheelers would be front ended, to be provided subsequently after the registration of the CSW. However, family benefits will be back-ended benefits that can be availed only after one or two years. A digital database would be maintained for tracking the number of people in the rolls, the number of days CSWs are engaged, and the frequency of their engagement. Family benefits can be availed only after one or two years based on whether the workers are regular or not.

10.4 IMPLEMENTATION OF GARIMA

10.4.1 Institutional arrangement

The institutional set-up for the scheme's execution at different levels, including decision making at the policy level, approval of necessary budgetary outlays, and review of the scheme's execution, is detailed below.

A technical support unit (TSU) has been set up at the state level with the support of the UMC[1] for effective implementation of the scheme. A state-level management information system (MIS) is being developed to record, collate and analyse the progress of implementation of the scheme (Figure 10.4).

10.4.2 Budget

A dedicated corpus fund of Rs 500 million has already been created by the state as a financial support to all the components ancillary and incidental to enhancing the safety and dignity of CSWs. The release and utilisation of the fund will be monitored by an exclusive administrative wing of the HUDD.

---

[1] Since 2005, UMC has been working closely with local governments by providing tailor-made solutions in professionalising urban management. With our understanding of adult learning pedagogies, capacity building and training, we work towards building individual and systemic capacities of cities.

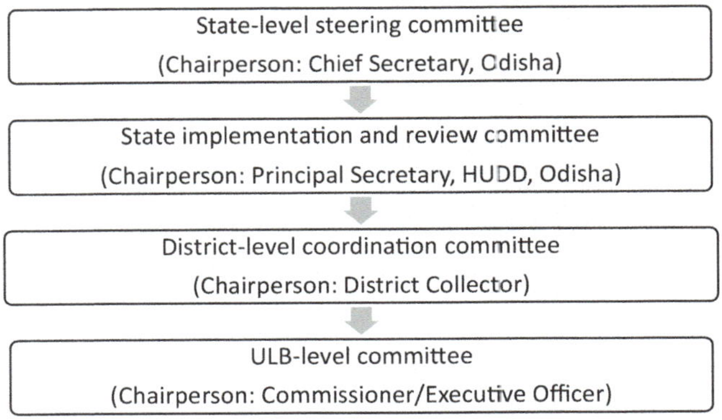

Figure 10.4 Institutional structure for implementing Garima. *Source*: Garima scheme document.

10.4.2.1 Roles and responsibilities of state departments

It is the state's responsibility to ensure safe working conditions for the CSWs. These workers cannot be engaged only through a private arrangement. The state has a responsibility to regulate the private sector and provide safe working conditions as well as extend social security benefits to CSWs and their families. Accordingly, state departments have been given specific roles and responsibilities relating to the programme (Figure 10.5).

10.5 PROGRESS ON THE IMPLEMENTATION OF THE SCHEME

- Enumeration of CSWs in all 115 ULBs of the state

In the beginning of 2021, an extensive enumeration process was taken up to identify and enroll all CSWs (both formal and informal) in six pilot cities (Bhubaneswar, Cuttack, Puri, Rourkela, Berhampur and Sambalpur) and later it has been scaled up to remaining 109 ULBs of the state. During the process, 13 977 surveys were conducted, during which 10 198 workers have been validated as CSWs so far. Swachha Sathis and Swachha Supervisors have been engaged in conducting the mobile application-based surveys on the ground. Given that most of the CSWs are employed by PSSOs under informal arrangements, there is a high turnover rate among staff. Consequently, regular enumeration of new CSWs is being ensured to address this issue effectively.

- Core sanitation work included in skilled and highly skilled categories of scheduled employment

Sanitation work was considered as an unskilled work in the scheduled employment. HUDD had a strong negotiation with Labour & ESI Department for revision in the categorisation and by a notification of Labour & ESI department, Categories of core sanitation work – such as (1) desludging of sanitation systems, (2) maintenance of the sewer network, (3) cleaning of inspection

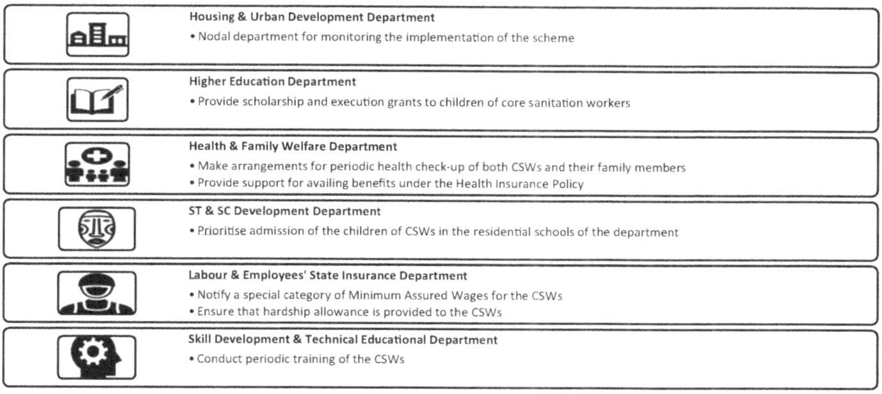

Figure 10.5 Roles and responsibilities of state departments. *Source*: Author's interpretation.

chambers by entering them, and (4) septic tank cleaning and maintenance – have been included in the highly skilled category and those performing such tasks designated as Core Sanitation Worker Grade-I. Other categories – such as (1) cleaning of drains, (2) cleaning, operation and maintenance of public and community toilets and (3) operation and management of STPs or FSTPs – have been included in the Skilled category and those doing such work designated as Core Sanitation Worker Grade-II. This has led to an increase in wages by 45% for Grade-1 and 27% for Grade-2 category of CSWs.

- Setting up of ERSUs and training and orientation of stakeholders

Emergency Response Sanitation Units (ERSUs) have been established in all 115 ULBs to enforce safety protocols during manual cleaning activities in situations where mechanical cleaning is not possible. While the formation of ERSUs was primarily guided by the advisory from the Government of India, they were originally intended to be established at the district level. However, considering practical implementation challenges, the state has deemed it more appropriate and feasible to constitute them at the ULB level.

Two distinct structures of Emergency Response Sanitation Units (ERSUs) have been implemented for sewered and non-sewered cities. In sewered cities, WATCO General Managers have been designated as Responsible Sanitation Authorities (RSAs), whereas in non-sewered cities, executive officers of the ULBs hold the RSA designation. Both WATCO and ULB officials, along with all ERSU members, have received training on the functionality of ERSUs.

- Procurement of prescribed PPE, safety devices and mechanical equipment

Standardisation of PPE, safety devices, and mechanical equipment is crucial in ensuring the safety and well-being of sanitation workers engaged in sewer and septic tank cleaning activities. In adherence to norms prescribed by the Central Public Health and Environmental Engineering Organization (CPHEEO), a comprehensive list of essential PPE, safety devices and equipment required for the complete mechanisation of all core sanitation services has been developed and shared with the ULBs. This standardised list serves as a guideline for ULBs in procurement processes, ensuring consistency and quality across all cities. CSWs now have access to the necessary protective gear, safety devices and mechanical equipment which is enhancing their safety, efficiency and dignity in carrying out the core sanitation jobs.

- Extensive training programme conducted for certification of CSWs as SEPs

To ensure the effective delivery of core sanitation services by trained and certified professionals, particularly in sewer and septic tank cleaning, an extensive 'Sewer Entry Professional' training programme has been initiated. Initially targeted at CSWs engaged in sewer cleaning and maintenance, the programme has now expanded to encompass all ULBs in the state. In addition to the OUA training centre in Bhubaneswar, two new regional training centres have been established in Rourkela and Sambalpur to maximise coverage within the stipulated timeline. These centres are equipped with simulation structures of sewers and septic tanks, alongside all necessary PPE, safety devices and mechanical equipment for hands-on practical training.

The training module is tailored to the educational levels of most CSWs, emphasising practical training. To date, 71 master trainers have been trained and empaneled with OUA and are providing training to CSWs to become 'Sewer Entry Professionals'. The Sewer Entry Professionals (SEP) initiative, initially focused on CSWs in sewer cleaning and maintenance, has now been expanded to cover all CSWs working in septic tank cleaning across the state. So far, 748 CSWs have been trained and certified as 'Sewer Entry Professionals'.

- Risk & Hardship Allowance for CSWs

Given the visible and invisible risks CSWs face daily during their work, a special allowance known as 'Risk & Hardship Allowance' has been extended to them. Sanitation workers engaged in sewer and septic tank cleaning received 15% of their daily wage as 'Risk & Hardship Allowance', while those working in Faecal Sludge Treatment Plants (FSTPs) receive 7.5%. This additional allowance not only provides them with a fair compensation for their contributions but also instills a sense of respect and recognition for their work.

- Procurement of protective uniform and wader suits with 'Garima' branding

A standard design for protective uniforms and full body wader suits has been developed in consultation with CSWs and some other stakeholders. The former is for use in the case of regular cleaning activities and the latter for emergency purposes which require entry into confined spaces. Water Corporation of Odisha (WATCO) has made an initial procurement of 2000 protective uniforms and 500 full body wader suits for CSWs engaged in the cities of Bhubaneswar, Cuttack and Puri, Sambalpur and Rourkela and now it has been scaled up to remaining cities as the vendors have been empaneled for the state.

- 'Garima Griha' rest lounges constructed for CSWs

Dedicated lounges called 'Garima Grihas' have been created for CSWs. These facilities provide them with adequate space to store belongings and protective equipment in personal lockers, a place to rest in air-conditioned halls, washing and drying areas, and a washroom. So far, 34 Garima Grihas have been set up across Bhubaneswar, Berhampur, Cuttack, Puri, Rourkela and Sambalpur

- Extension of disability support and illness allowance to CSWs

As CSWs are exposed to hazardous work environments and are prone to accidents and injuries at the workplace, the state has issued a notification for inclusion of all eligible CSWs under the Employee Compensation Act, 1923, and Employees' State Insurance Act, 1948. The notification specifies that those CSWs not eligible for inclusion under these Acts shall be provided with the same benefits through Garima.

- Reduction of work hours

Since the services rendered by the CSWs are hazardous and arduous in nature and it is not desirable to work for eight hours while exposed to toxic gases in harmful, unsafe and accident-prone conditions, the government has notified that a Grade-I CSW shall work for six hours per day (excluding half an hour of rest) and shall be entitled to full wages as notified under the Minimum Wages Act, 1948, despite the reduced working hours.

- Extension of social security benefits to CSWs

As a part of the extension of social security benefits to CSWs and their families, a special drive has been initiated for enrolment of all identified and validated CSWs under the Biju Swasthya Kalyan Yojana (BSKY). So far, around 4572 workers have been enrolled under the scheme across the six pilot cities.

However, the state has noted that approximately 30% of identified CSWs with Garima Identity cards were deemed ineligible for enrolment under BSKY due to lacking the required ration card, which was the sole eligibility criterion. A proposal was submitted to include CSWs in BSKY irrespective of possessing a ration card, with the Garima ID serving as the sole eligibility requirement, and it was approved. The process has been initiated with the health department to extend the benefits to CSWs and their families.

- Elevation of experienced CSWs as 'Garima Supervisor'

In the recent initiative, existing experienced CSWs having reading and writing abilities have been engaged as 'Garima Supervisors' for the purpose of monitoring Sewer & Septic Tank (both mechanical & manual) cleaning keeping in view of their experience and intimate understanding of ground realities involved in various sanitation operations. The idea is to leverage their core competency and elevate them to create a pool of resources who could be deployed to monitor hazardous cleaning operations.

Under this pilot initiative, WATCO has identified 22 nos of CSWs/SEPs from Bhubaneswar, Cuttack and Puri based on the predetermined eligibility criteria. However, after training, 15 SEPs have been found eligible in terms of experience, training and reading and writing skills. These successful SEPs have been deployed for monitoring purposes in the jurisdiction of Bhubaneswar and Cuttack Municipal Corporations and Puri Municipality.

10.6 TEETH TO LEGISLATION

Public health and sanitation are included in the State List of the Seventh Schedule of the Constitution of India, which means any teeth to legislation relating to these needs to come from state governments. Given this fact, the ability of state and local governments to translate the Centre's policy direction has proven to be a missing element in addressing this issue. Given the essential nature of the work in question, local governments and municipal bodies have found it a tricky course to navigate.

The Garima scheme is unique in the legislative and policy landscape of India. It represents a standout effort on the part of one state government to address an issue on which the Centre has so far provided broad direction through its legislation. In this sense, it serves as a bridge between the Centre's intentions and the flesh-and-bones system that is required on the ground to ensure that the intentions do not just remain on paper but are translated into reality.

It is in this context that the Garima scheme stands out as a successful and progressive framework against hazardous core sanitation work. So much so that the Garima scheme has inspired the Ministry of Social Justice, GoI which recently introduced a programme called "NAMASTE"., National Human Rights Commission (NHRC) of India has also recommended to all states to adopt the Garima scheme.

10.7 AWARDS AND RECOGNITION

Garima has been recognised as a 'Best Practice' by the National Human Rights Commission, highlighting its impactful contributions to social welfare. In 2023, Odisha was awarded the best in 'Skilling in Sanitation' by ISC – FICCI, demonstrating the state's commitment to enhancing sanitation skills among its workforces. Additionally, Odisha received the 'Special Recognition Award for Government' by ISC – FICCI in 2021, further cementing its leadership in public sector initiatives. The Garima scheme has also garnered international acclaim, receiving accolades from organisations such as the Sanitation Workers' Forum (SWF) and the Water and Engineering Development Centre (WEDC), reflecting its significant global recognition and influence.

Manual Scavenging

Source: HUDD, GoO.

Mechanised desludging

Source: HUDD, GoO.

Mechanised desludging

Source: HUDD, GoO.

Garima Hall

Source: HUDD, GoO.

doi: 10.2166/9781789064926_0175

Chapter 11

Employment safety net for the urban poor: UWEI/Mukta scheme

ABSTRACT

To address the rising unemployment and vulnerability of informal labour and migrants, the Odisha government launched Urban Wage Employment Initiative across all cities of Odisha. This chapter focuses on the genesis of the scheme, salient features and its implementation in Odisha, discussing the innovative governance model in which community-based organisations have emerged as the fourth tier of governance that helps in creation as well as maintenance of community infrastructure in the urban areas.

Keywords: community driven, employment opportunity, women empowerment, infrastructure creation, urban poor, unorganised sector, migrant workers, informal workers

11.1 SAILING PAST A GRAVE STORM

The national lockdown imposed on 25th March 2020 due to COVID-19 was soon followed by visuals of desperate migrant workers scrambling to return to their hometowns or villages. These images from some of India's largest cities eclipsed the second part of the story: what happened to the workers once they reached home.

The second part of the story has been of critical importance in a state like Odisha. The reverse migration of workers was a deluge, given that an estimated 1 million Odisha residents lived and worked in other parts of the country. A survey of reverse migration in six states showed that the incomes of the returning migrants in Odisha fell by a devastating 93%[1] as compared to their previous workplace. This, combined with a 23.8% national unemployment rate

[1] https://www.livemint.com/news/india/70-reverse-migrants-ready-to-return-as-income-falls-sharply-11602235295765.html

in the last week of March 2020,[2] created a huge livelihood crisis, aggravating the vulnerability of the economically weaker sections of people.

The pandemic-induced lockdown hit the urban workforce the hardest, with factories, offices and establishments remaining closed. News reports convey that many of those who returned to their rural homes did so with the hope of earning some form of income locally through the MGNREGA and other such schemes. Those who returned home to urban centres, however, had no such safety net.

COVID 19 pandemic pushed most informal workers in Odisha deeper into poverty, affecting jobs and livelihood. As a result, about one-fourth of the urban population faced severe consequences, especially with reduced working hours, decreased wages and abrupt layoffs. In addition, around 1 million migrant workers returned to Odisha from other states due to the closedown of economic activities. At this critical juncture, the Housing & Urban Development Department (HUDD) of the Government of Odisha (GoO) swiftly responded by launching the Urban Wage Employment Initiative (UWEI) on 18 April 2020. Conscious efforts to undertake labour-intensive projects helped avert profound economic crisis by creating rapid mass employment opportunities for the urban poor, informal and migrant labourers who were rendered unemployed and vulnerable in the face of the pandemic. It also helped enhance city resilience by creating climate-sensitive durable assets across the city. Urban areas were at the forefront of the ongoing pandemic, with the urban poor bearing a disproportionate burden, further marginalised due to abrupt loss of livelihood. MUKTA, launched in the face of the pandemic, is poised to mitigate such adverse impacts while building economic resilience among poor urban communities. It leveraged the strong community networks of Self-Help Groups and Slum Dweller Associations, building capacities, thereby strengthening the fourth tier of governance – a new institutional mechanism for creating city social resilience. In addition, its convergence with Jaga Mission (a state-wide slum upgradation initiative) transformed the poor urban settlements by upgrading basic civic and social infrastructure and integrating them into the city fabric. Such transformative actions are aided by handholding support from City Governments, standardised design specifications and estimates, and periodic grievance redressal in all 115 cities across Odisha.

11.2 GENESIS OF THE INITIATIVE

The idea of an employment initiative for urban areas is not new. An urban wage employment programme was first proposed during the 8th five-year plan, when poverty alleviation was a central national goal. The Swarna Jayanti Shahari Rozgar Yojana (SJSRY) was launched in 1997 after subsuming three earlier urban poverty alleviation schemes, namely Urban Basic Services for the Poor (UBSP), Nehru Rojgar Yojana (NRY) and Prime Minister's Integrated Urban Poverty Eradication Programme (PMI UPEP).[3] This scheme had two components:

[2] https://thewire.in/economy/india-unemployment-lockdown-cmie

[3] https://employmentbankwb.gov.in/sjsry.php

one focused on generating self-employment and the other on generating wage employment. Later, it was replaced by the National Urban Livelihood Mission (NULM) in 2013. While these schemes recognised the vital role of community-level organisations in creating wage employment opportunities for the urban poor, they were underfunded and provided limited flexibility to implementers. And none of them were employment guarantee schemes. In 2020, Kerala started an urban wage employment guarantee scheme called Ayyankali Urban Employment Guarantee Scheme (AUEGS) on the lines of MGNREGA.

With millions of people left without any livelihood across towns and cities of India, the impact of COVID-19 and the economic lockdown was indeed unprecedented, especially in the unorganised sector. The mass exodus of migrants due to loss of income and uncertainty about the future forced governments to think of providing welfare to those severely impacted, which precipitated the revival of interest in wage employment schemes in urban areas. Odisha took the lead in resolving to start an urban wage employment programme during the lockdown.

Quick action for immediate relief

A few days after the lockdown was announced, when the scale of the livelihood crisis became evident, the MoHUA called a meeting with all state governments to know their views on ways to provide relief to those who had lost their livelihood. Many state secretaries suggested the launching of a scheme to provide wage employment opportunities as an immediate safety net for migrant workers and urban poor across the country who were facing a livelihood crisis.

Realising the constraints in launching any new scheme at the national level and consequent delay or inaction, we decided to go ahead with a wage employment scheme for the urban poor in Odisha. After intensive deliberations of the core team in the department, the outline of the scheme was ready within a week.

Fully cognisant of the gravity of the situation, the Odisha Chief Minister's Office was quick to give us the go-ahead over a couple of telephone discussions and exchange of proposals and notes during the lockdown. The funding for the new scheme was leveraged from the existing UNNATI scheme thus saving valuable time on getting clearances and approvals from the Finance Department (including its empowered committee which scrutinises and recommends new schemes). As a result of this, the new scheme could be launched swiftly – within three weeks of the start of national lockdown.

– G. Mathi Vathanan

The scheme was launched with the short-term objective of providing relief during the pandemic as well as the long-term objective of ensuring livelihood protection for low-skilled urban poor. Along with livelihood protection this programme has two equally important objectives: creation of community assets by identifying and implementing labour-intensive projects in the urban local bodies (ULBs) and strengthening the fourth tier of governance by engaging community-based organisations such as Mission Shakti Women Self Help Group, Area Level Federations (ALF), City Level Federations (CLF) and Slum Dwellers' Associations (SDAs) as implementing partners in the execution of these projects.

Following Odisha, other states like Jharkhand,[4] Himachal Pradesh,[5] Tamil Nadu and Rajasthan have since launched similar initiatives. The launch of urban wage employment schemes in Odisha and other states has initiated a national-level policy discussion on different aspects of such programmes. In September 2020, Jean Dreze, a noted welfare economist who had played a vital role in the formulation of MGNERGA, proposed an MGNERGA-like scheme for urban areas, called DUET (Decentralised Urban Employment and Training),[6] which advocates guaranteed employment like MGNERGA and opportunities for skill upgrading. Another variant of the scheme, proposed by the Centre for Sustainable Employment at the Azim Premji University,[7] argues that urban wage employment initiatives can provide livelihood security for informal workers and act as a strategy for creating new infrastructure and maintaining old infrastructure. It too highlights the critical role of community organisations and ULBs in the effective implementation of such programmes.

The initial experience of UWEI in Odisha has shown that such schemes play an important role in reducing the vulnerability and enhancing resilience of the urban poor during economic shocks caused by disasters like a pandemic. Among all the states, the unemployment rate in Odisha reached the pre-pandemic level fastest.[8] According to a survey of eight states, Odisha had the lowest share of reverse migrants who wanted to go back. The scheme has been able to employ female workers who are worst affected due to economic shocks. It has also successfully created community infrastructure in slums and other marginalised neighbourhoods, which will reduce the vulnerability of the poor during a pandemic or similar natural disaster in the future.

11.3 APPROACH AND COVERAGE

UWEI was simultaneously implemented across all 115 ULBs in Odisha primarily to provide wage employment for urban poor, including migrant workers, unemployed youths and other vulnerable groups like disabled, transgender and socially backward castes. Most of them are employed in the unorganised sector and hence have little job security and low wages. In time

---

[4] Mukhya Mantri Shramik Yojna (English translation).

[5] Mukhya Mantri Shahri Ajeevika Guarantee Yojna (MMSAGY).

[6] https://frontline.thehindu.com/economy/interview-jean-dreze-duet-is-just-an-idea-now-india-unemployment-crisis/article65674725.ece

[7] https://cse.azimpremjiuniversity.edu.in/duet-expand-to-include-social-protection-for-informal-workers/

[8] Odisha Economic Survey, 2020–2021.

of crisis these groups are often disproportionately impacted; therefore, this programme emphasises livelihood security to build social resilience among those working in the unorganised sector.

The livelihood situation of urban poor in the unorganised sector, especially those engaged in home-based work, street vending or self-employment, is dependent on their access to community infrastructure. Improved public infrastructure in their places of residence can improve their productivity by connecting them with better opportunities. For some, public infrastructure like community centres can act as workplaces for SHGs or other such groups. These places can also be used for other community activities like organising meetings, discussions and other community events.

Creation and maintenance of community infrastructure in the past has been challenging because of low community involvement. The contractor-led implementation model is driven by profit and often neglects community requirements. Involvement of the community in the creation and maintenance of infrastructure could be an alternative to contractor-based models. The experience of the Odisha government in involving SHGs and other community groups shows that it decreases corruption, enhances the quality of implementation and results in increased ownership of the community as illustrated in Figure 11.1. This encouraged the state government to involve the community in all stages of project implementation. Participation of the community as the fourth tier of governance is key to the success of welfare schemes like MUKTA.

Under UWEI, multipurpose community centres such as Parichay for slums and Mission Shakti Gruha (serving as a common roof for collective action of SHGs) for non-slum areas have been built which can be used for various community activities including economic (self-employment) activities of individuals and groups. They can also develop into public spaces. Maintenance of natural resources through open space development around water bodies is also being taken up under this initiative. Other related projects like rainwater harvesting structures and stormwater drainage are also being developed and maintained. In the first year of the UWEI, special focus was on beautification of the city which led to the creation of opportunities for artisans and painters.

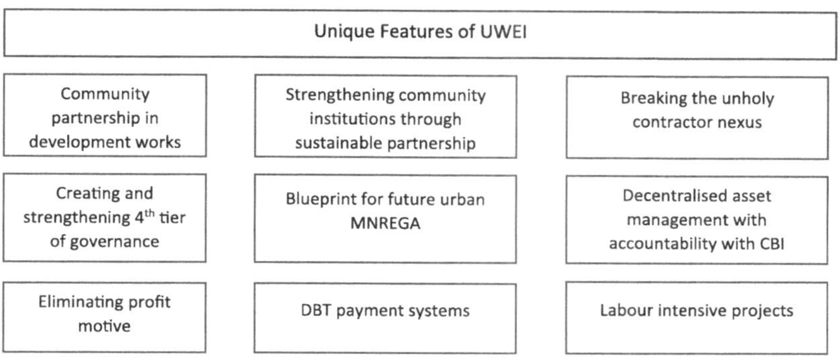

Figure 11.1 Unique features of UWEI. *Source*: adapted from UWEI guideline.

11.4 PROJECT COMPONENTS

11.4.1 Stakeholders

The successful implementation of a large and critical scheme like UWEI depends on the active participation of several stakeholders. UWEI has identified two kinds of stakeholders: primary and secondary. The primary stakeholders are the direct beneficiaries: the urban poor, unemployed youth, local community and vulnerable groups. The secondary stakeholders are those involved in scheme implementation, such as the functionaries of ULBs, SHG federations, SDAs and Mission Shakti Groups. The primary and secondary stakeholders need to synergise for the initiative to be successful. For this purpose, the scheme has a well-defined institutional architecture with clearly laid-out procedures and responsibilities.

11.4.2 Budget and permissible works

Initially, UWEI was launched with an outlay of Rs 1000 million for six months and was later extended for another six months with additional Rs 1000 million. The ongoing pandemic further underscored the long-felt void of an urban wage employment programme as a counterpart to and in line with India's long-standing rural employment guarantee programme, MGNREGA. Given the widespread positive impact of UWEI and the bottom-up demand for a livelihood safety net for the urban poor, UWEI transitioned into a well-embedded regular State-funded scheme, MUKTA, within a year with a five-fold increase in financial allocation.

The organic growth of MUKTA as a scheme with support across the political spectrum adds to the initiative's sustainability. Moreover, MUKTA's innovative models and approaches for the creation of participatory community assets have scripted a new paradigm of planning and implementation of urban infrastructure development programmes which is now followed for executing projects under other schemes and programmes such as 15th Finance Commission Grants, Jaga Mission and UNNATI, among others. The state missions of UNNATI and JAGA share a common main objective of creating inclusive infrastructure in the cities. UNNATI was launched by the government to promote community-based assets in cities. JAGA Mission aims to give formal rights to slum dwellers and upgrade basic infrastructure in slums. Keeping these objectives in mind, UWEI identified six broad categories of community infrastructure and climate resilient works to be carried out: building of *Parichaya* centres for slums and *Mission Shakti Gruhas* for non-slum areas, storm water drainage, green cover and beautification, rainwater harvesting and open space development around water bodies.

In the initial period of the UWEI, which was UWEI the focus was on projects related to building of Parichay centres, Mission Shakti Gruhas and beautification drives involving local artisans. This year government is focusing on developing rainwater harvesting structures in the cities. All the works implemented under UWEI projects were designed to be labour intensive and ensure a ratio of 60:40 between labour and material costs, to maximise wage employment possibilities.

Multiplier effects of UWEI across programmes

The UWEI programme is significant not only in being timely from the COVID-19 perspective and the reverse migration that Odisha faced, but also because of its flexibility in aligning with the objectives of other state government programmes such as the JAGA Mission. Soon after UWEI was launched in September 2020, the JAGA Mission also started pursuing the objective of universal slum upgrading across all cities in the state. The funding available through UWEI helped the ULBs to create community assets as discussed in the JAGA Mission Chapter 7. Services and infrastructure such as paver block roads and drains particularly benefitted from the UWEI programme with SHGs and SDAs involved in providing labour and creating these assets in each slum. Therefore, UWEI also contributed to the JAGA Mission's objective of slum upgrading. The slum upgrading programme itself led to improved water, sanitation and electricity creating delisted slums and leading to slum-free cities; therefore, a virtuous cycle began with the financing available with the UWEI programme. Figure 47 demonstrates how the circularity of these programmes has been designed in the context of Odisha.

– Shubhagato Dasgupta

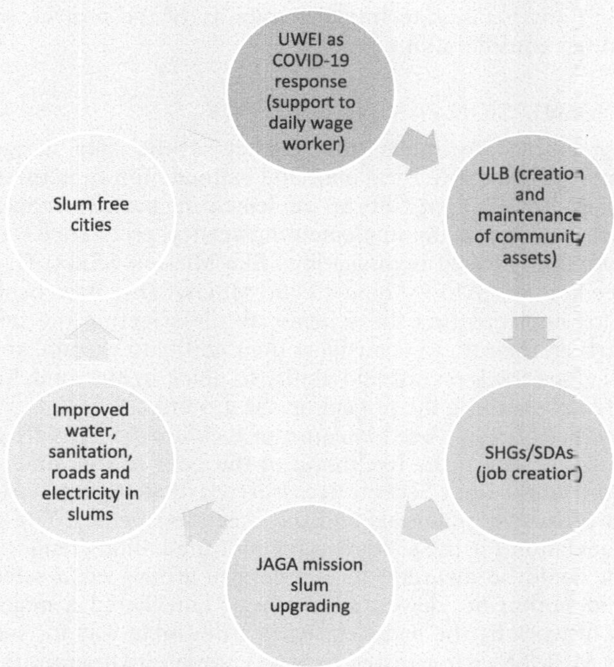

11.5 DECENTRALISED INSTITUTIONAL SET-UP

11.5.1 Ward-level committee

Ward-level committees became active during COVID-19 and played an important role in managing the situation. The committee has six members which include the Ward Officer – the member convener who is appointed by the municipal government – Assistant Engineer AE/Junior Engineer JE in-charge of the ward, President/Secretary of that ward, President/Secretary of one or two Mission Shakti SHGs of the ward, President/Secretary of SDAs/slum-level committee for each slum, and Swachha Sathis/Supervisors as decided by the Municipal Commissioner (MC)/Executive Officer (EO). The ward-level committee is responsible for identifying a list of projects after ascertaining their technical feasibility through AE/JE. This allows the community members to prioritise projects according to their needs.

11.5.2 ULB-level committee

A ULB-level committee is set up with the MC or EO as chairperson. The other members of the committee are City Engineer/Municipal Engineer, Health Officer, Child Development Project Officer, Nodal Community Organiser and City Mission Management Unit (CMMU) managers, President/Secretary of the SHG federation, President/Secretary of the SDA federation, or erstwhile SDA president, and any other official as decided by MC/EO. The role of the ULB-level committee is to finalise the list of projects based on the need for wage employment in any locality, technical feasibility of the project, scope of the project and other considerations.

11.6 IMPLEMENTATION PARTNER/AGENCY

Another key innovation under the project is the role assigned to an implementation partner. After the plan and estimate for an activity or project are prepared, the MC/EO appoints an implementing partner or agency (IP/IA) for the project. In selecting an implementing agency, preference is given to the SDA and community-based organisations like Mission Shakti Groups or the Area Level Federation (ALF). About 15 000 MSGs/SDAs have been partnered with so far for implementing the scheme. While selecting the implementing agency or partner, the MC/EO considers their ability to execute and supervise the work, previous track record and ability to abide by the guidelines. The IA takes care of implementing the project on the ground.

With the offices being closed because of lockdown and swift government action required to secure the livelihood of the poor at this time, substantial deviation in the Works Code became necessary, particularly related to the public procurement process of materials and the executing agency. The government approved a deviation for the scheme to bypass the tedious tendering process, so that work could be awarded to community groups on a selection basis. However, this cannot be viewed adversely or considered a major deviation as execution of work by the beneficiaries is a desirable way for such projects. Works under MUKTA are entrusted to local community organisations without a profit margin who are the ultimate beneficiaries.

11.7 DIRECT BENEFIT TRANSFER

Quick payment of wages is vital for the success of the programme. The daily wage to the labourers is pegged at Rs 326/day for unskilled workers, Rs 348 for semi-skilled workers and Rs 398 per day for skilled workers. These rates are revised as per the minimum wages in the Labour Act. As the community organisations do not have the capacity to advance daily wages and the labourers are dependent on regular payments, the scheme adopted the direct benefit transfer (DBT) payment system to transfer wages to the workers' accounts every week. The process of approving the payments was simplified and streamlined so that the payment process was smooth for the labourers as well as the community organisations.

11.8 PROJECT IMPLEMENTATION

The implementation of such schemes requires close coordination among a range of stakeholders including functionaries of the ULBs, community members and organisations. The government of Odisha issued several Guidelines and SoPs for smooth functioning of the scheme, clearly setting out the role of each stakeholder to ease implementation. The implementation of the UWEI/MUKTA scheme involves the following stages.

11.8.1 Preparing and finalising the list of projects

Identification and preparation of the list of projects is the first step of UWEI. The ward-level committee is responsible for identifying the projects to be undertaken in their respective areas. The projects are chosen from the list of permissible works as defined in the guidelines. The selection is based on the priority for a particular area as decided by the ward committee. The AE/JE responsible for each ward then conducts the financial feasibility analysis of the project and prepare estimates for it.

The list of projects thus finalised by the ward committee is sent to the ULB-level committee for administrative approval as illustrated in Figure 11.2. The committee approves the projects based on the need for wage employment in that area along with technical feasibility of the work. As part of transparency, the final list of projects is displayed on the notice boards of the ULB, ward office

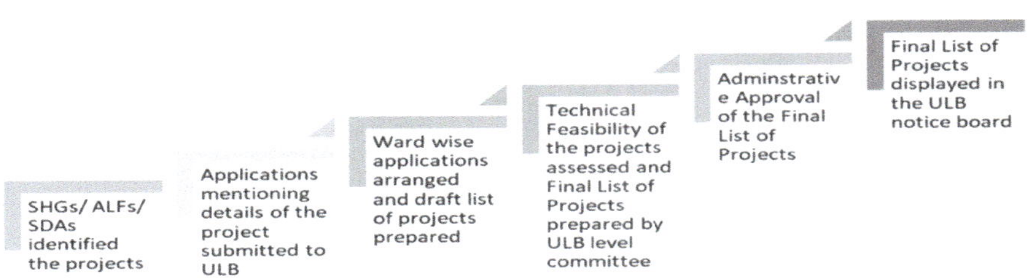

Figure 11.2 Finalising the list of projects. *Source*: adapted from the UWEI SoP.

and Child Development Project Officer (CDPO) office and the ULB website for knowledge of the public.

11.8.2 Issuance of work order and start of work

Following preparation of the plan and estimates, and administrative approval of these, the MC/EO selects an MSG/SDA as IA/IP. SDAs are given priority as IPs for projects and only in cases where the SDA does not exist are MSG/ALF from slums engaged as IP/IA. To ensure equitable distribution, not more than one project is awarded to any SDA/MSG.

Depending on the cost of the project, the work order is issued to either the SHG/SDA/ALF or AE/JE. If the project cost is 1 million or less, the work order is issued to the IP/IA; if it is above 1 million, the work order is issued to the AE/JE (however, it can also be given to the SDA based on the requirement of the ULB). Upon receiving the work order, material required for the work is procured. In cases where the work order is issued to the AE/JE, he/she will procure the material and hand it over to the IA/IP as illustrated in Figure 11.3. Based on the project estimate, the muster roll is issued to the IP/IA responsible for the maintenance of the muster roll, which forms the basis for payment to the workers. More recently based on the experience of the implementation of the scheme, now the projects are fully executed end to end by the community partners (IA) and no longer through IP route.

11.8.3 Monitoring the progress of work

The monitoring and progress of work under the scheme is the responsibility of the IP/IA. They ensure the maintenance of all necessary project records, registers, documents and so on manually or digitally as prescribed by the guidelines. The AE/JE is responsible for monitoring to ensure that expenditure

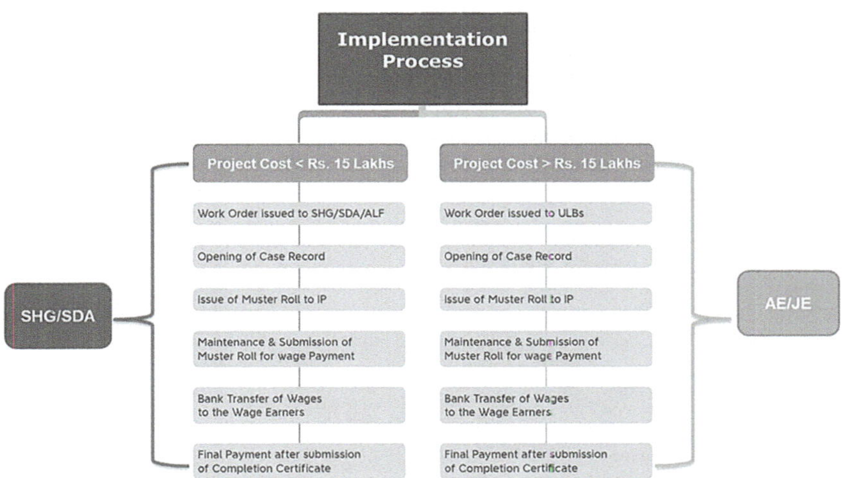

Figure 11.3 Implementation process. *Source*: adapted from UWEI SoP.

is within the prescribed limit and the technical standards described in the plan are maintained. All work-related measurements are recorded in the measurement book. Measurements are a must for each task to avoid unpaid work. The measurements are taken weekly immediately after the closure of weekly muster. In case of any deviation from the plan the AE/JE suggests corrective measures, allowing for mid-project corrections if required.

11.8.4 Payment of wages

As per the MUKTA Guideline, weekly payments are to be made to workers directly from the scheme account. The wage rate is as per the rate fixed by the Labour Department and Employees' State Insurance Corporation (ESIC). All payments under this scheme are made through DBT; there are no cash transactions. The IA is paid supervision charges of 7.5% of the total estimated cost of the work if the work order is issued to it. However, if the work order is issued in favour of AE/JE, then the IA/IP is paid 7.5% of the total wage component of the work as supervision charges. For participation in meetings, the ULB pays sitting fees out of its own funds: Rs 250 per meeting to the President and Secretary of MSG/SDA, and Rs 500 per meeting to the President and Secretary of ALF.

11.9 AMA POKHARI SCHEME: AN EXAMPLE OF HOW UWEI HELPED ACHIEVE BROADER ENVIRONMENTAL GOALS

The Government of Odisha's vision of making urban Odisha water secure and ecologically diverse has identified the need to rejuvenate waterbodies in urban areas. Hence, the Urban Waterbody Rejuvenation Program has been initiated by the Department of Housing & Urban Development, Government of Odisha. More than 2000 waterbodies have been identified in all urban areas of the state to be mapped and rejuvenated in a planned and sustainable way. A new scheme has been launched to rejuvenate Urban Water Bodies in a natural and sustainable manner, with the help of Mission Shakti Groups (MSGs). This scheme has been named as 'Ama Pokhari'.

The key objectives of the programme encompass several interventions in line with MUKTA at enhancing urban water bodies. First, it focuses on the meticulous mapping of these water bodies, ensuring proper documentation to facilitate effective management and preservation. Additionally, the programme seeks to develop city-specific action plans for the rejuvenation of water bodies, tailored to address the unique challenges and opportunities within each urban area. A significant component of the programme is the capacity building of multiple stakeholders, including ULBs, WRU and MSGs. This involves equipping them with the necessary knowledge and skills to manage water resources effectively. Furthermore, the programme aims to conduct a large-scale public awareness campaign to highlight the values, functions and ecosystem services provided by urban water bodies. This will be achieved through IEC activities, utilising multiple media platforms to reach a broad audience and foster a deeper understanding and appreciation of these vital resources.

Ama Pokhari lays a new approach towards Water Body Rejuvenation that revolves around enhancing the quality of water, sustaining the rejuvenation with the help of community. The key principles are:

- Rejuvenation with minimal to nil use of concrete.
- Focus on water quality improvement and soil–water interface (ancient methods).
- Campaigning under the 'Mukhya Mantri Karma Tatpara Abhiyan' (MUKTA) as part of the Urban Livelihood Mission. Implementation will be driven by Mission Shakti Partners.
- Supervision of Water Bodies Rejuvenation works during execution phase by Interns.
- Captivating In-House Drawings/Estimates/DPRs.
- Focus on generating more workman hours instead of machine hours.
- Focus on nature-based solutions (ecologically) making the rejuvenation process relatively economical.

11.9.1 Making community the champion behind waterbodies rejuvenation

The impact of Ama Pokhari is clearly visible beyond the water's edge. It recognises the immense potential within communities, particularly women, to be the driving force behind this transformation. Mission Shakti Self-Help Groups (SHGs), led by women, are at the forefront of the rejuvenation efforts. They are not just passive participants; they are the core implementing partners who are de-weeding, desilting and ensuring the plantation activities, while being the dedicated guardians of these revitalised water bodies. When the programme was introduced about a year and half back, there were so many eye-brows raised, so many questions were thrown at the Department; 'how can house wives as part of Self Help Groups understand the technicalities of waterbody rejuvenation?' 'How can the embankment survive without stone pitching or heavy concrete structure?'

However, this programme along with environmental restoration has become a thriving model of social empowerment. The 'Ama Pokhari' programme equips women with the knowledge, skills and resources to actively participate in shaping their city's future. It provides them with a platform to showcase their leadership and become agents of positive change. Also Ama Pokhari scheme for the very first time in the country gave the responsibility on the youth who are fresh graduates from local colleges the opportunity to supervise the rejuvenation process having joined as interns with the Department. Their support made it easy for officials in urban local bodies to scale up the activities and set high targets.

Traditionally, waterbody rejuvenation projects tend to often be contractor-driven, with a focus on quick, concrete solutions that may not always be ecologically sensitive. Ama Pokhari challenges this paradigm. By prioritising community involvement and nature-based solutions, it fosters a sense of ownership and responsibility among residents. When communities are invested in the process, they become the long-term stewards of these revitalised water bodies ensuring their continued health and preventing them from slipping back into neglect.

11.9.2 Scale and speed

HUDD, Odisha in almost all its innovative endeavours have very diligently worked towards timely completion without compromising the quality and scaling up any pilot programme for the benefit of the public throughout the state. For Ama pokhari too, the success of pilot waterbodies made us realise the programme is not only doable through community partnership but also it is scalable as the expenditure per acre has been 20–25 lakh which is one of the lowest in the country. A team of efficient, superannuated engineers who have worked in the state for substantial number years and understand the nuances of topography and weather of Odisha were willing to team up as Waterbodies Rejuvenation Unit who designed, scrutinised, prepared estimation for the DPRs and in a short span of six months an incredible number of 407 DPRs were ready and approved by the Government to be taken up across the state. For ease of monitoring a dashboard comes handy for the officers on the field as well as in the headquarters.

11.9.3 A ripple effect of change: from local to global

Ama Pokhari's impact has a ripple effect that extends outwards, transforming the city as a whole. The programme serves as a beacon of hope, not just for Odisha, but for cities across India and beyond. It demonstrates that ecological restoration is possible, that nature-based solutions can be effective, and that community participation is the cornerstone of long-term success.

Ama Pokhari is more than just a project; it's a movement for change. It's a testament to the power of collective action, a celebration of nature's resilience, and a recognition of the vital role women play in shaping a sustainable future. As each stagnant pond is transformed, Ama Pokhari paints the picture of a city not merely surviving, but thriving, in harmony with nature.

11.10 IMPACT AND RECOGNITION

UWEI provided rapid relief to reduce livelihood vulnerabilities not only of the urban poor in Odisha, but also of migrant workers stuck in the state due to the pandemic and migrant workers returning to the state due to distress in-migration from surrounding regions. Further, with COVID-19 exacerbating gender inequality, UWEI addressed the gender impacts of the pandemic by (a) putting women SHGs at the centre of the planning, decision-making and implementation of the initiative, (b) including gender perspectives in all data and coordination mechanisms and (c) providing safe and secure working environments for women workers. The large-scale engagement with community groups to create community assets has enhanced the capacity of Mission Shakti Groups and Slum Dwellers' Associations for undertaking such community asset development projects. This has resulted in a steady demand for uptake of such community asset development projects by community organisations and urban poor communities.

UWEI was able to provide relief to many people during and after the lockdown by providing wage work as illustrated in Figure 11.4. About 0.6 million urban poor got employment through the initiative so far, and more than

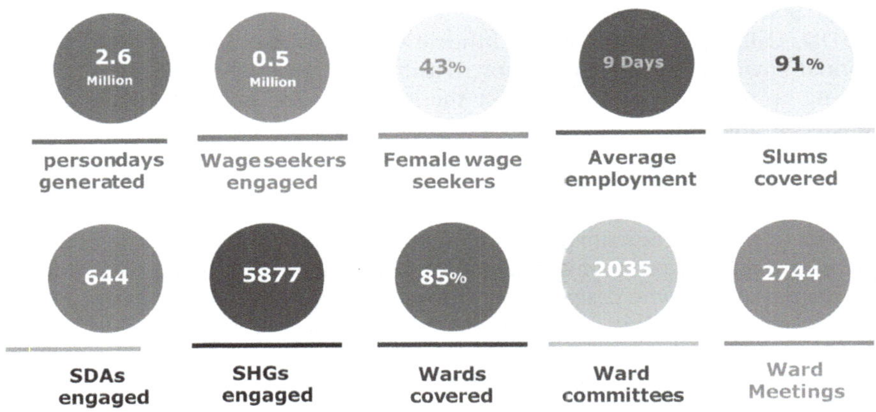

Figure 11.4 Snapshot of impact (needs revision based on figures mentioned below in text). *Source*: government of Odisha.

7.6 million person-days of work have been generated. Women are 40% of the workers employed under this scheme, showing that it has been able to provide relief to women workers who are disproportionately impacted by such disasters.

Around 56 500 projects with an investment of Rs 13 000 million have been implemented across 1906 wards out of 2055 wards (97% coverage) across 115 ULBs. As much as 95% of slums have been covered under the initiative. More than 18 500 CBOs (MSG and SDAs) have partnered in the implementation of the scheme. More than Rs 320 million has been paid to SHGs and SDAs as supervision charges. The proportion of wards that have benefited through the creation and upgradation of infrastructure has been 85%. Ward level committees have been formed in all the 2055 wards and conducted ward meetings, ensuring active participation of the community in the implementation process.

Conversion of UWEI to MUKTA

The success of the UWEI initiative encouraged the state government to convert it into a full-fledged scheme called Mukhyamantri Karma Tatpara Abhiyan (MUKTA) with a larger financial outlay and more robust implementation structure. The Government of Odisha has earmarked Rs 10 000 million under MUKTA for providing meaningful employment to the urban poor and creation of sustainable community assets across 115 ULBs of the state. The MUKTA scheme has been designed based on the learning from the implementation of UWEI as well as other national and international case studies of employment guarantee schemes. Intensive consultations were carried out with field-level officials to understand the challenges and opportunities in upgrading the initiative to a full-fledged scheme. Consultations were also done with other stakeholders like IAs/IPs who provided valuable inputs on scheme implementation.

Capacity building of IPs/IAs and ULB officials and timely payment of wages emerged as key issues in scheme implementation. The revised guidelines of the MUKTA scheme have paid special attention to ensure timely wage payment and smooth clearance of payment to IPs/IAs by leveraging information technology. Adequate provisions for capacity building of stakeholders have also been made under the MUKTA scheme.

Under MUKTA, the government of Odisha has achieved important breakthroughs in programmes such as the Ama Pokhari scheme discussed above as well as 'Catch the rain: where it falls and when it falls' campaign through construction of Rainwater Harvesting Structures (RWHSs) for in-situ water conservation and groundwater recharge in open spaces as per land availability and technical feasibility. The campaign was executed in mission mode before the onset of the south-west monsoon in 111 ULBs of the state covering 1990 urban wards, with the active participation of women SHGs and SDAs. All these RWHSs are low-cost structures specially designed based on hydro-geological parameters prevailing in the state. A total of 37 235 RWHs were constructed with an investment of Rs 1500 million within a record time of three months of the launch of the campaign, to national acclaim. The plan was to construct 30 000 RWHSs by 2022, spread across all 2055 wards in 115 ULBs. However, the government surpassed the target by establishing almost 37 000 RWHSs, along with 598 Open Space Developments like parks and playgrounds.

The Ama Pokhari programme, despite being in its early stages, has already made impressive progress in rejuvenating urban water bodies in Odisha. To date, over 2000 water bodies have been identified across 115 ULBs in the state. Comprehensive surveys have been completed for 800 of these water bodies, and DPRs have been prepared for 416, with nearly 407 officially sanctioned at a total value of 310 crores. The programme has also mobilised 1943 Mission Shakti Self-Help Groups (SHGs) to actively participate in rejuvenation efforts. Furthermore, 150 interns have been onboarded to provide additional support and expertise. Work has commenced on rejuvenating 371 water bodies, with four already completely restored and many others in advanced stages of progress. These achievements signify substantial strides towards a healthier and more vibrant future for Odisha's urban water bodies, showcasing the power of community and nature in driving transformative initiatives and inspiring other cities to follow suit.

The core materials used for constructing community assets under MUKTA are fly-ash bricks, morrum, crusher dust, laterite sandstone, cement and sand. Odisha is a hub for thermal power plants, resulting in an abundant supply of fly-ash bricks. These bricks are environmentally friendly as their prime constituent is ash generated as a waste product from the thermal power plants. As these plants are situated within the state, it further reduces the carbon footprint due to reduced transportation distances while closing the recycling loop. In addition, using paver blocks for the roads ensures the percolation of water that recharges the water table. This small yet essential intervention has the potential to reduce flood risks in these settlements and build a resilient community.

MUKTA focuses on creating ward-level basic amenities and infrastructure, leading to equitable development of various areas within the city. The Government of Odisha has always sought to achieve speed and scale without compromising on quality and transparency in the implementation of MUKTA. With a budgetary provision of Rs 10 000 million, MUKTA is poised to achieve a new benchmark in the urban livelihood horizon in India. The community-centric innovations in the scheme have strengthened the fourth tier of governance by building local leadership with a focus on women empowerment to further synergise long-term sustainable development around the urban community in Odisha.

The Government of India has appreciated and applauded the efforts of the Odisha government and recommended Odisha's MUKTA model to other states for replication. The states of Jharkhand and Tamil Nadu with a combined urban population of 43 million have announced and rolled out schemes on the line of Odisha's MUKTA for the creation of employment opportunities for the urban poor.

This initiative has also received global recognition, winning two international awards – Medal of Honour and Online Popular City Award – at the prestigious 2020 Guangzhou International Award for Urban Innovation. Entries were selected from 273 initiatives in 175 cities and 60 countries across the world. Furthermore, MUKTA has been awarded the 2021–2022 WRI Ross Center Prize for Cities on 29 June at the 11th World Urban Forum in Katowice, Poland. The prize recognises MUKTA's swift action to create mass employment and infrastructure during the COVID-19 pandemic.

Key takeaways of District Collectors involved in programme implementation

MUKTA is a scheme where the public is a critical stakeholder in its implementation. With successful completion of projects, communities are further encouraged to take up more projects.

It is a public-driven scheme – driven by the people's demand for work and local assets – and implemented by the people themselves with constant support from the city government.

– Chakravarty Singh Rathore

11.11 A STITCH IN TIME

UWEI came into being in the context of the COVID-19 pandemic. During the initial months-long total lockdown and subsequent less stringent ones, while the organised sector quickly fell into work-from-home patterns, this was not an option for most of the unorganised sector, which makes up the bulk of the Indian economy and accounts for about 90% of Odisha's economy. This meant that the biggest impact was on a section of the population that was already financially strained and vulnerable. This lent a sense of urgency to the process of designing and rolling out of the scheme, which was achieved in about three weeks.

Once the immediate objective of providing some economic relief and support to the urban poor was achieved, the state managed to successfully raise the impact profile of the initiative by reorienting it to focus on long-standing challenges. The expansion of UWEI into MUKTA may be seen as a systemic attempt to address the twin factors of poverty and migration. Odisha is one of the largest source states for domestic migration, and MUKTA could address this challenge at least in part.

Strengths of the programme

I believe Odisha's UWEI initiative has clearly demonstrated that if a public work programme is well designed to provide sensitive social protection, it can significantly reduce urban vulnerability.

There has been continuous strengthening and capacity building of the fourth tier of governance by empowering community groups in areas of planning, execution and monitoring; adequate incentives have been provided to the public to build overall community resilience. UWEI has demonstrated the potential for community involvement in both service delivery and contributing to building human capital.

- Harsh Kothari, then Consultant to SCI FI

The ward-level committees help in expanding community collaboration for schemes and projects to raise resilience among urban poor against economic shocks or challenges like climate change continued success of MUKTA will depend in no small part on capacity-building efforts. This will not only help ensure that there is a trained and available workforce for the projects, but also contribute to raising the skill profiles of workers, potentially leading to larger earning potential. Further training of the SHGs and SDAs involved in the projects will also help build up the social capital of leadership and managerial skill at the grassroots level.

Odisha's ever-evolving experience with UWEI/MUKTA is now available as a blueprint for other state governments and the Centre to pick up and adapt to their own contexts. The intent and political will of the state leadership has been a key ingredient in the scheme's success. This helped Odisha stage one of the fastest recoveries from the economic impact of the lockdown amongst Indian states and return to pre-pandemic levels of unemployment in a matter of months.

Mass exodus

Source: HUDD, GoO.

Paved walking path

Source: HUDD, GoO.

Open space development

Source: HUDD, GoO.

Wall painting

Source: HUDD, GoO.

doi: 10.2166/9781789064926_0195

Chapter 12

Other flagship initiatives of HUDD, Government of Odisha

ABSTRACT

This chapter highlights other flagship initiatives launched by the Government of Odisha to further the transformation of urban Odisha. It also discusses the Odisha Urban Academy, which is driving the change agenda by disseminating new practice-based knowledge and designing and delivering innovative training solutions for various stakeholders. These efforts aim to support and facilitate transformative changes across the entire urban landscape.

Keywords: Digital door numbering, streetlight, capacity building, smart LED lighting, skill development

12.1 BACKGROUND

In recent years, Odisha has embarked on pathbreaking urban transformative changes. The state has conceptualised and executed numerous pioneering initiatives such as 24 × 7 water supply/drink from tap (DFT), decentralised and non-sewered faecal sludge and septage management (FSSM), decentralised solid waste management (SWM) and GARIMA. These initiatives have set benchmarks and attracted the attention of policymakers and practitioners at the national and international levels. Recently, Odisha has also utilised digital technology through the digital door numbering project to ensure essential public services, easier communication and improved logistics. This technique allows individuals and their homes to be traced more conveniently.

Additionally, the Government of Odisha has launched the smart light-emitting diode (LED) public street lighting project, leading to better illumination, increased vehicular and pedestrian safety during night hours, energy savings through modern and robust infrastructure and a reduction in greenhouse gas emissions and carbon footprint. These evidence-based leading practices across the urban landscape, coupled with community-led and inclusive schemes, provided the impetus to scale up and replicate these initiatives across the state within a 2–3-year time horizon. Recognising the need to further build capacities at the individual and institutional levels and to strengthen the training and skilling ecosystems, Odisha established the Odisha Urban Academy (OUA). The OUA aims to transfer learning and widely disseminate practitioner-focused solutions to a broader audience.

12.2 DIGITAL DOOR NUMBERING

Generation of resources for the creation of infrastructure and providing quality services in urban local bodies (ULBs) is a huge challenge and therefore, the Central and State Finance Commissions, through their recommendations have advised the state governments to prioritise the subject. The Government of India has also been issuing regular instructions indicating the goal and timeline to enhance the resources of ULBs and reduce their dependency on government grants. An increase in the pace and extent of urbanisation leads to a corresponding growth in the properties and this in turn opens a perfect opportunity for ULBs to augment their property tax revenue in a systematic and time-bound manner. Odisha, having 17% of urban population, is a less urbanised state in comparison to other states. This further broadens the scope of ULBs to generate revenue by utilising progressive strategies and advanced technology.

12.2.1 Rationale for launching digital door numbering

Analysis of the current scenario revealed that around 550 000 properties have been subjected to assessment, which indicated that another 5 lakhs (approximately) more properties can be added to the existing tax net. As the survey and valuation of un-assessed properties is a massive exercise, it requires the assistance of a qualified agency who can bring all such properties into a common fold on geographic information system (GIS) platform enabling ULBs to plan for assessment of property tax with subsequent annual increment as suggested in the Reforms Agenda. The digital platform consisting of all such properties will ensure that there is uniformity in assessment and collection with regular focus on annual enhancement. Similarly, the properties captured along with utility services provided in different streets will also enable ULBs in assessing the service charges and in identifying the gaps, if any.

The database can also be utilised for integration with several other platforms and secure a single-window delivery of services. The integration of multiple systems can also enhance the efficiency of accounting and financial management.

With this backdrop, as a part of a unique urban intervention, this department started a new and unique initiative called digital door and street numbering (DDN), which was proposed to be implemented in 112 ULBs of the state except for three other ULBs. This will provide numbers to all the households and the streets in the ULBs.

12.2.2 About the programme

The DDN solution for all the dwellings in the jurisdiction of 112 ULBs of Odisha was designed to have a geo-spatial solution, depicting an alphanumeric smart address code for each property/establishment in respective ULBs with the information of main road, sub-road/lane, landmarks, building and information on the number of floors. This will ensure the benefits of providing navigation details and address location services to the dwellings/properties within the jurisdiction of respective ULBs and help see spatial depiction of information related to electricity, water connection, property tax and fee collection management.

The agency was proposed to be mandated to conduct a door-to-door survey using a GPS-enabled system to identify the properties/establishment/vacant land. The following information was required to be collected during the survey in an application to be developed by the agency for the survey:

- Full address including owner name, legacy property/holding number (if available) and landmarks.
- Street name, ward number and pin code.
- Picture/image of building/structure elevation – photo image of the front view of the building with respect to the road to be captured as a part of the work; and
- Building/dwelling type (commercial/residential/mixed/apartment/vacant/school/offices/temples/community halls/other) with construction type and so on.

The agency was required to affix a weather-resistant, rust-free metal board with a printed unique digital door number for each dwelling after code generation and collecting the required information at the holding level. The size and the content of the DDN plate were supposed to be specified by the department. Every number should be linked with Google Maps so that location searching can be performed by using that number. The agency will be responsible for providing all necessary support and authorisation as and when required. Based on the IDs, the system should be able to provide address information to one or more navigation, routing and GIS systems. At the same time, navigation should not be limited to specific platforms but should work with all available products on the market.

With the above objectives in view, the department had floated a request for proposal (RFP) specially curated for the purpose and selected an agency, that is, M/s GIS Consortium India Pvt. Ltd., through a fair and transparent bidding process. The work order in this regard was issued and the contract was signed with the agency. In the first phase, a pilot survey was completed in Khurda municipality and the project work commenced at 111 other ULBs shortly thereafter. The period for implementation of this project was 9 months followed by an operation and maintenance (O&M) period of 1 year.

Based on pilot town experience, standard operating procedures (SOPs) had been prepared and survey work was scaled up in other 111 ULBs from 30th July

2021. The following activities were devised and detailed SOPs were issued with the following points:

- Wider awareness is to be created among the citizens of the ULBs regarding kick start of the DDN project. The people will be communicated about the initiative, survey process, its benefits and usage. Public awareness will have to be made by posting hoardings, announcement during garbage collection, advertisement in newspapers and so on.
- Kick off meeting with ULBs and collection of inputs such as ward map for marking of actual ward boundary, ULB property/holding data, authorisation letter for survey and so on.
- Starting of ward-wise door-to-door survey through a mobile app in on-line and off-line modes and geo-tagging of properties/holdings during field survey and capturing of photo for each property/holding.
- Once survey is completed and DDN is generated for all holdings, DDN records will have to be shared with ULB officials for verification on the field and on satellite map through DDN Web application to check the overall coverage.
- After the generation of DDN for all the holdings, DDN plates will have to be installed on all the properties.
- Unique DDN number will have to be integrated with property/holding tax, water connection, electricity consumer number and so on for seamless delivery of services.

12.2.3 Progress on the ground

So far, a door-to-door survey based on the finalised SOP has been completed with the help of a real-time mobile application. Survey teams have visited ward-wise door-to-door and collected response-based information according to the questionnaire developed in the mobile app and geo-tagging all the properties and syncing of the same on the server through a desktop web application has been completed. As of today, a DDN survey has been completed in all 112 ULBs and 14.50 lakhs DDN unique numbers have been generated.

After completion of survey and finalised the DDN numbers, DDN door plates have been finalised by the department and installation work is in progress and completed in 93 ULBs. DDN plates have already been affixed at 1 081 238 out of 1 450 000 properties and 6670 out of 30 000 streets and it has been targeted to affix all the DDN plates by 30th October 2024.

Based on the common attributes, DDN database has also been integrated with the property tax database in respect of 105 ULBs. Besides, the data have also been successfully leveraged for the unique initiatives of the department relating to grey-water management and other sanitation activities. There have been several learnings in the process, which will be of immense value for others who intend to replicate the initiative at scale.

12.2.4 Way forward

Even as the initiative is nearing completion, the prospect of utilising the valuable geo-spatial data and textual information captured in the process appears to be baffling. Whether in routine affairs of the ULBs or in matters relating to analytics, the data promises to be an incredible input. The initiative was started with a simple objective of augmenting revenue of the ULBs, but it has opened such windows, which were hitherto unthinkable.

The DDN data can be leveraged in the following ways:

- Augmenting the revenue of the ULBs by rationalising the property tax assessment.
- Real-time updating of the property register by virtue of the geo-tagged properties.
- Instant updating of arrears and other details pertaining to properties through the mobile application.
- Streamlining door-to-door garbage collection and monitoring of the entire SWM value chain.
- Water supply management and billing.
- Real-time updating of water supply connection and distribution details.
- Unified revenue collection using QR codes and DDN.
- Utilities/assets information management.
- Application and renewal of trade licenses.
- Other departments may utilise the DDN data for the following activities:
- Electricity billing and distribution.
- Distribution of ration and other emergency aids.
- Fire services.
- Cooking gas connection and distribution.
- Telecommunication infrastructure and so on.
- Grievance redressal of the citizens based on real-time and authentic data.

Hence, DDN can address many of the daunting challenges faced by the city managers.

12.3 PUBLIC STREET LIGHTING

ULBs are mandated to provide basic minimum services for a decent life of urban citizens. One of the most important services is public street lighting, and a thrust area to provide adequate illumination for safety for pedestrians and vehicular traffic. The Government of Odisha through ULBs under Housing and Urban Development Department (HUDD) has initiated installation of smart energy-efficient LED luminaires across the ULBs in the state.

The overall situation of the public street lighting system was quite challenging. The public street lighting system until 2014 with high energy-consuming conventional lighting and infrastructure was outdated and inefficient, resulting in inadequate illumination on the streets impacting safety and security with risks such as accidents, crime incidents and decreased visibility for motorists and pedestrians alike. Moreover, high energy bills were the norm for all ULBs. Most of the ULBs were unaware of energy efficiency, saving of energy and handling of energy systems in a comprehensive manner.

To tide over the issues, the HUDD decided to cover all public streets in the urban areas of all ULBs through energy-efficient smart LED public street lighting infrastructure, thereby leading to better illumination, and increased vehicular and pedestrian safety during night hours, energy saving through modern and robust infrastructure and reduction in greenhouse gas and carbon footprint.

It started with an experiment of a retrofitting project in 2014 in Bhubaneswar Municipal Corporation for replacement of conventional lamps with energy-efficient LED luminaires and upgrading of infrastructure. Subsequently, the Odisha Street Lighting Project was taken up in five municipal corporations in the public–private partnership (PPP) mode in July 2016.

To cover all other ULBs, the HUDD had selected another implementing agency in August 2017 for replacement of conventional lamps with energy-efficient LED luminaires and upgrading of infrastructure.

Furthermore, due to addition of new areas and expansion of ULBs, the requirement for public street lighting had been steadily increasing and most ULBs had requested for installation of additional public street lighting infrastructure in greenfield/uncovered areas for the safety and security of the public at large. The HUDD initiated the Greenfield LED Street Lighting Project in the PPP mode in 113 ULBs in December 2018.

The projects have been implemented by the various agencies with the O&M period of 7 years. Independent engineers were engaged for verification and validation of the installed quantities with the technical specifications as mentioned in the contractual documents/agreements.

To continue the O&M activities and carry out replacement and installation in stretches wherever required in the ULBs, new agencies have recently been appointed for installation and O&M for 5 years period for both greenfield and retrofitting stretches.

The HUDD also appointed a separate project management unit in Odisha Urban Infrastructure Development Fund (OUIDF) involving various technical, financial and other professionals for monitoring the progress and operation of the projects and other activities.

The present system of management of streetlights through retrofitting and greenfield systems has provided a new lease of life and a way of management of street lights where in, it is controlled through a computerised control monitoring system (CCMS) and is monitored through a dashboard. The existing system makes the agencies responsible for keeping the streetlights glowing as it is linked to their performance. The payment system to the distribution companies (DISCOMs) and to the agencies has been streamlined. The repair and maintenance of the street lighting system by the agencies with service level benchmarks is a new chapter in the management of streetlights in the state of Odisha.

As it is a dynamic thing, efforts are afoot to provide services to newer urban areas as per its growth with due diligence and maintain the old system with efficiency.

Currently, around 496 326 number of energy-efficient LED luminaires has been replaced and installed in 115 ULBs having total project cost of Rs. 702 crores. The current ongoing project in ULBs for the greenfield and retrofitting stretches is estimated to entail a capital cost of Rs. 480 crores. The projects have been funded by the Government of Odisha, OUIDF through KfW-Germany and ULBs.

12.4 ODISHA URBAN ACADEMY

Odisha in the recent years embarked on pathbreaking urban transformative changes. The state has conceptualised and executed many pioneering initiatives such as 24 × 7 water supply/DFT, decentralised and non-sewered FSSM, decentralised SWM and GARIMA thereby setting a benchmark and attracting attention of policy makers, and practitioners at the national and international levels. Such evidence-based leading practices across the urban landscape in general and community-led and inclusive schemes representing the people-centric approach in particular, not only provided the impetus to scale up and replicate it in a 2–3 years' time horizon across the state but also acted as a trigger for Ministry of Housing and Urban Affairs (MoHUA) of the Government of India to issue advisory to replicate such home-grown model schemes in Atal Mission for Rejuvenation and Urban Transformation (AMRUT) cities across the country.

12.4.1 Rationale for setting up Odisha Water Academy

Having recognised the imperative to further build the capacities at the individual and institutional levels as well as strengthen the training and skilling ecosystems to transfer the learning and widely disseminate the practitioners-focused solutions to a wider audience, Odisha Water Academy (OWA) was set up as a Centre of Excellence in 2021 under the aegis of the HUDD, Government of Odisha. The academy became operational in July 2021 under a training licit mandate to provide training and capacity building support to the personnel working at managerial and worker levels in the drinking water and wastewater sectors.

During its formative stages from July 2021 to March 2023, the OWA emerged as a leading anchoring institution of state and national importance to have trained 2196 professionals by catering the wide-ranging demands of diverse stakeholders from the utilities/cities in the state as well as other states in the country and neighbouring countries. Incidentally, Central Public Health & Environmental Engineering Organization under MoHUA has recognised the OWA as a leading training institution in the WASH domain.

12.4.2 Evolution from the OWA to OUA

Meanwhile, the state in the process of experimentation and innovation has implemented new models of urban development and governance that stand out as the role model for other states. To meet the emerging needs for an institution to provide training and upgrade the skill sets of stakeholders at all levels (including policy makers, officials, elected representatives, community partners and contractors), the Government of Odisha decided to expand the scope of the OWA by changing its name to 'Odisha Urban Academy' in 2023. The academy aims to turn into a preferred destination to meet the capacity building needs not only of the state but also to address the emerging needs of other states as well as countries interested in adopting the Odisha models.

Having recognised the catalytic role of the OUA in capacity building and institutionalisation of sustainable urban development as well as taking a cue from women self-help group (SHG)-led participatory models of Odisha in

24 × 7 water supply/DFT, MoHUA, Government of India, roped in OUA as a knowledge partner to develop knowledge products and information education and communication materials to launch a nation-wide 'Women for Water-Water Jal Diwali' campaign during 7–9 November 2023, that witnessed participation of 14 000 women in water governance all over the country. MoHUA is equally keen to partner with the OUA to emulate the Odisha model of developing master trainers to train AMRUT Mitra (women SHGs) on water quality testing, analysis and monitoring. Incidentally, the OUA under the guidance of the HUDD has recently partnered with IIT Madras to conduct a training of master trainers on water quality testing, analysis and monitoring. These master trainers will train lab technicians/chemists, sample collectors, Jal Sathis and O&M staff of water treatment plants (WTPs) under a state-wide capacity building programme under AMRUT 2.0 in the next 2 years in a phased manner.

The academy is currently a sprawling learning hub located within the sanitation complex surrounded by sludge treatment plant, faecal sludge treatment plant and wealth centre at Basuaghai under Sisupalgarh gram panchayat. Spread across 32 000 square feet, the academy is a state-of-the-art training centre and equipped with two conference halls, five lecture halls/ smart classrooms, a library and cafeteria. A residential complex is constructed to accommodate 62 persons.

The academy is registered under the Societies Registration Act, 1860. The management of the affairs of the academy is entrusted to the 12-members governing body under the Chairmanship of Additional Chief Secretary, HUDD. It is under the administrative control of the State Urban Development Agency, HUDD.

12.4.3 OUA's vision and mission

The vision and mission of the OUA is to be a leading institution in training, research and advocacy to achieve sustainable and inclusive urban development by building capacities of individuals and organisations for urban transformation. It ensures integrity, innovation, inclusivity, excellence, collaboration and sustainability as its values. The broad thematic areas and some of the flagship offerings of the OUA are as follows:

- Planning, design and implementation of 24 × 7 water supply/DFT Mission.
- O&M of WTPs.
- Water quality testing, analysis and monitoring.
- Planning, design and implementation of rural water supply projects.
- FSSM and urban–rural convergence in sanitation.
- Grey-water management.
- Integrated municipal SWM.
- GARIMA mission (to ensure occupational health, safety and dignity of core sanitation workers).
- Capacity building for community partners (women SHGs/transgender (TG) SHGs/slum dwellers association).
- Gender sensitisation to ensure diversity, equity, inclusion and accessibility at the workplace.

- Leadership development and team building.
- Digitalisation of urban service delivery (SUJOG and MUKTA soft).
- Capacity building for The Urban Learning Internship Program (TULIP) interns to develop a resource pool of young future leaders in urban development.

The academy during the period since inception has organised capacity building, exposure visit and institutional training for the state-/national-level participants. Besides, delegates from the neighbouring countries (Nepal and Bangladesh) and African countries (Rwanda and Ethiopia) have taken part in the programme on 24 × 7 water supply and wastewater/SWM.

The faculties are largely from the HUDD, utilities and knowledge partners. Many of them are retired government officers who are domain experts. Recently, efforts have been made to source faculties from reputed institutions such as IITs, XIM University and corporates to bring the latest cutting-edge knowledge and best practices on table. The OUA has signed MoUs with national and international institutions such as NIUA, USAID, ASCI, UMC and WASH I for collaboration in capacity building and action research. Recently, TATA Trust has shown keen interest to identify leading mentoring institutions for institutional collaboration.

The academy right from the beginning has been operating in self-financing mode. To cover its administrative overheads, it charges Rs 25 000 per day or 25% of the total training charges, whichever is higher.

12.4.4 Impact

Odisha has made significant strides in urban development by focusing on capacity building and skill enhancement across various thematic areas. A total of 6073 professionals have been trained, covering diverse sectors crucial for sustainable urban management. Specifically, 1423 participants were trained in drinking water, water quality and RWSS, whereas 286 public health engineering organisation engineers received specialised training. Additionally, 465 individuals were educated in FSSM and wastewater, and 466 TULIP interns focused on urban sanitation. The GARIMA initiative witnessed the participation of 910 professionals, and 209 individuals were trained under SUJOG. Community engagement was emphasised with 1130 community partners, including TG, Jal Sathis and SHGs, receiving training. Furthermore, 1001 professionals were trained in MUKTA, urban planning and governance, with 74 participants focusing on grey-water management. Thirty-two IDCO junior estimators (PH) and 77 individuals attended various workshops and seminars, collectively enhancing the state's capacity to implement and manage urban infrastructure and services effectively.

Odisha's comprehensive training initiatives have attracted a diverse group of participants, enhancing the state's urban development capacity. Of the 6073 professionals trained, a significant majority, 87% (5311 participants), were from within the state, highlighting a strong local commitment to capacity building. National participants accounted for 10% (612 individuals), indicating a broader interest in Odisha's pioneering urban practices. Additionally,

Table 12.1 Overview of the professionals trained from July 2021 to May 2024.

	Year				
	2021–2022	2022–2023	2023–2024	2024–2025 (Up to 15th May)	Total
No. of programmes	22	42	92	11	167
No. of professionals trained	583	1613	3494	383	6073
No. of days	55	117	215.5	35	422.5

3% (150 participants) were international, reflecting global recognition and the international relevance of Odisha's training programmes. This diverse participation underscores the wide-reaching impact and appeal of Odisha's efforts in urban development and capacity building (Table 12.1).

In line with the vision and mission, the academy in the next 2–3 years would like to reposition itself as an institution of national and international importance to organise high-impact training in urban development to saturate the capacity building needs for all the ULBs/utilities within the state and AMRUT cities. In keeping with the digitalisation of learning ecosystems, it would like to introduce and launch certified online and/or hybrid courses for different target audience in collaboration with national- and international-level institutions.

doi: 10.2166/9781789064926_0205

Chapter 13

Community partnerships towards building a fourth tier of urban governance

ABSTRACT

Odisha's urban development department has demonstrated that partnerships between community-based institutions and city governments can deliver transformative urban change. Slum dwellers' associations (SDAs) as area-based community structures are being capacitated to go a step ahead of the model of self-help groups and other urban poor community groups actively participating in the development and governance process. SDAs are being positioned to provide a governance level, referred to as the fourth-tier functioning under the third tier which is the urban local body (ULB), furthering the spirit of decentralised governance under the Constitution. In this manner, SDA and its relationship with ULBs is transforming the urban governance framework in the state. This chapter aims to bring out the key enabling decisions by the government that have and are contributing to building and empowering the fourth tier in cities through capacity building, institutional strengthening and economic and financial empowerment.

Keywords: Decentralisation, community-based organisations, capacity building, community resilience, infrastructure ownership

13.1 BACKGROUND

The principle of decentralisation was enshrined in the Indian Constitution in 1992 through the 74th Constitutional Amendment, which formally recognised urban local bodies (ULBs) as the third tier of government. It mandated that the state governments transfer to local governments a set of specified functions under the 12th Schedule, such as urban planning, including town planning; the provision of water; public health and sanitation and solid waste management (SWM).[1]

[1] https://www.tandfonline.com/doi/full/10.1080/07352166.2016.1271614.

Odisha has widely demonstrated that partnership between citizens, municipalities and the state can deliver transformative urban change in the state not only at the conceptual level but also in operation. Mission Shakti groups, transgender groups, ragpickers' groups, among others are now active partners of ULBs and participate in most government programmes. Urban poor communities are actively involved in the development process across the 115 cities in the state.

Building on the constitutional decentralisation mandate, and further advancing its spirit, the Odisha government has sought to foster a fourth tier of governance – in the form of community-based organisations (CBOs) called slum dwellers' associations (SDAs) – and thereby develop sustainable community resilience and ensure socio-economic empowerment. This novel and innovative governance approach, by creating space for the fourth tier of governance also strengthens the third tier that is the city governance structure itself in a mutually reinforcing manner so that the spirit of the constitutional amendment can be realised and the service levels.

By stimulating and fostering the capacity of SDAs the government aims to ensure that the urban poor who have been marginalised for a long time get a strong voice and encouraged to develop a greater sense of ownership and responsibility towards the city. This unique and novel effort is being developed alongside building a strong ULB. This cross-cutting effort is defining and contributing to most of the programmes discussed in the earlier chapters. This fresh approach is a defining characteristic of the Odisha model wherein the building of the capacities of the urban poor has wider benefits, in addition to being providing access to basic infrastructures and services, communities are becoming more cohesive and more resilient and better placed to confront economic and social challenges if new challenges show up in the future.

13.2 REINFORCING THE THIRD TIER OF GOVERNANCE AS A NECESSARY CONDITION

The 1992 Constitutional amendment provided an impetus to the idea that with evolving urbanism and changing socio-political realities, the role of the national government must change from controlling to guiding. It recognised that decentralisation of governance is key to upholding the principles of accountability, transparency and good governance. The devolution of power makes the local governments increasingly accountable for improved service delivery and operation and maintenance (O&M) of urban infrastructure and services.

Coming out of an era where state governments were mainly responsible for urban development most regions have taken some to operationalise decentralisation and face the critical challenges that need to be solved to transition to a decentralised urban governance model. Often, the reforms must transcend sectoral policies and foster cooperation between different actors and stakeholders relevant to local contexts, ensuring a balanced distribution of capacities and resources. Taking on this challenge as the urbanisation process picks up pace in Odisha with a clear understanding that under-equipped local governments can sometimes compromise planning processes, financing and transparency and accountability, the Department of Housing and Urban Development (HUDD) has invested in improving the capacity of ULB personnel vital for effective implementation of plans, and efficient urban management and service delivery.

The Government of Odisha has ensured that the state's model of urban governance conforms to the 74th Constitutional Amendment Act (CAA) and takes in inputs from the UN-HABITAT's Global Campaign on Urban Governance by empowering local governments and increasing their capacity for participatory urban planning and decision making. The robust decentralisation process in its urban governance framework has ensured that the ULBs involve citizens and communities in planning and implementation of developmental schemes in cities, which has steadily contributed to improving the capacities of the city government and improved the quality of services provided by the third tier of governance. It has built on the Government of India programmes since the 74 CAA, such as Jawaharlal Nehru National Urban Renewal Mission and Atal Mission for Rejuvenation and Urban Transformation (AMRUT) among others, Odisha's urban programmes have consistently promoted decentralisation and participation. Furthermore, they have also created opportunities for furthering decentralisation at every opportunity.

The genesis of the decentralisation at scale in HUDD programmes in the urban areas of the state was first experimented in the septage management programme in 2017. In Odisha, when the initial faecal sludge and septage management (FSSM) projects under AMRUT were undertaken in the nine class 1 cities, much of the construction was directly overseen by the Odisha Water Supply and Sewerage Board (OWSSB) as a state-level agency: it carried out the design, planning and construction oversight through private contractors. However, to scale it up to cover the other 115 ULBs, the approach was changed to further decentralisation with ULBs in the lead, thereby also building the capacity of the third tier of governance. Hence, each ULB was given the authority to contract out and oversee local construction with the oversight of the OWSSB. The devolution empowered the ULBs not only to give out contracts, but also oversee the quality and understand the technical issues involved. Furthermore, community participation capacities of ULBs were also built as the HUDD encouraged them to engage with self-help groups (SHGs) in construction of public toilets, providing O&M services for around 150 community toilets and public toilets and operating cesspool vehicles. Communities were also encouraged to play vigilant roles on indiscriminate disposal of septage/manual cleaning in wards.

Another example is the evolution of the Jaga mission. During the Jaga mission – phase 1, when land rights were being handed out, there was an important requirement for state control on data and plans through the Jaga mission office. ULBs and District Collectors were involved in validating the data and ensuring that the Urban Area Slum Rehabilitation and Redevelopment Committees (UASRRCs) functioned well. The data were generated by agencies working with the central Jaga mission office at the state level. However, when Jaga 2 began, it was seen as an opportunity for the Jaga mission to further decentralise and push the agenda of the fourth tier of government to carry out slum upgrading across all the cities. The ULBs played a central role in coordinating and engaging with SHGs and SDAs and thus implementing the slum upgrading programme. Hence, ULB capacity was built in undertaking these local jobs in partnership with CBOs. This exercise as discussed in the earlier chapters on the mission built the capacity of ULBs alongside the capacities of the SDAs and SHGs and established clear frameworks for participation and cooperation. The Jaga mission established that the decentralisation route is often the most effective for scaling up infrastructure and services within a participatory framework. This programme exemplifies the government's decentralised approach: it enabled decision making at the ULB and ward levels, established the UASRRCs and brought the SDA into the forefront as competent community-based structure capable of playing a strong role in urban governance.

The state's reforms in the field of SWM also demonstrate a staunch resolve to empower and build the capacity of urban local governments through decentralisation, planning and decision making, while simultaneously involving community groups. Towards achieving the state's agenda for SWM, the government issued a standard operating procedure (SOP) for decentralised SWM to all ULBs in 2019 that laid out a step-by-step process – providing the ULBs with technical knowledge and capacity – to successfully implement SWM as discussed in a chapter on SWM. Furthermore, the ULBs as the third tier of governance also gained from the experience of invoking the participation of Swachha Sathis, community volunteers who are responsible for anchoring the decentralised and community centric SWM services.

A key lesson from Odisha's model is that decentralisation of decision making for ULBs and building their capacities is a necessary condition for scaling up infrastructure and improving the quality of services in cities. The next sections further discuss that while an empowered, well-governed third tier is a necessary condition for improving services, deepening participation and dignity of poor communities is also a critical factor to ensure city resilience and the sustainability of high-quality services and the further evolution of the urban governance framework through the establishment of the fourth tier of government.

13.3 GENESIS OF STRUCTURED COMMUNITY PARTICIPATION IN PROGRAMMES AND ITS IMPORTANCE IN THE URBAN DEVELOPMENT MODEL OF THE STATE

Traditional planning and programme implementation regimes have often ignored the voiceless, such as the poor and marginalised groups, and decision making is primarily under political and bureaucratic control. Giving teeth to this often discussed but under used idea of community participation in government programmes recognising the indispensable role of local communities in urban development the Government of Odisha has implemented programmes that adopt a participatory approach by engaging the communities, thus making for a more need- and demand-based line of action. Active engagement of communities in the planning and implementation process increases the likelihood of the development plans being more closely matched with the needs, interests and expectations of the stakeholders, while also increasing the financial feasibility of projects.

In the 1990s, the SHG model was functioning in Kerala and Andhra Pradesh, and these were presented as best practices. Learning from them, the Government of Odisha started mobilising women SHGs during 1997–1999 in the district of Ganjam. These SHGs pertained to rural areas and were running financially remunerative enterprises. The experiences gained from scaling up the SHG model in the district have had a significant impact on formulating ideas for pursuing the implementation of the fourth tier of governance.

The HUDD started adopting and expanding the community participation approach as its schemes started becoming large and when large-scale infrastructure construction was not the main challenge anymore and much of the challenges were in service delivery and operation and maintenance of the infrastructure. The HUDD made a strategic decision that instead of relying on local petty contractors for service delivery and operation and maintenance they would encourage ULBs to work with local community groups, such as SHGs. This strategic decision required the schemes to be designed in such a way that the capacities of the ULBs were built to engage with communities and other stakeholders on a continuous way. This approach also meant that the capacities of SHGs were to be built across many activities in urban operations management, be it cesspool operation, faecal sludge treatment plant services, SWM, bill collection, among many others. Consistently, each programme of the HUDD that came to be designed and implemented from 2018 onwards built in these aspects in the scheme design as the earlier chapters bring out.

The decision to involve community groups in urban service delivery as against contractors was driven based on several factors. Firstly, the idea was that the involvement of community groups would build greater shared ownership of the infrastructure and help build wider attachment to the city itself. The second was based on the poor experience of involvement of contractors in public service delivery in the past. Another factor was that through the participation of local community groups greater accountability, transparency and better service quality could be ensured. This strategic choice was driven by a host of other reasons too, which included the limited experience of the local private sector in these urban services activities, to build on community involvement to help create new livelihoods for the poor and build their socio-economic resilience.

The SWM programme saw the largest number of SHGs in urban services. In 2019 when the Odisha government started the SWM programme it already had successful working relations with SHGs in eight ULBs, in running the septage plants built under the FSSM programme. The government decided to make the SWM programme SHG-centric too, with no involvement of contractors; it designed simple technologies and leveraged the locally available resources and capacities in the form of SHGs to manage SWM. The OWSSB selected the SHGs and entrusted the ULBs with capacity building of the SHGs, followed by contract signing for carrying out the tasks assigned to them. The Swachha Sathis as community volunteers function as change agents to cause behavioural changes at the household level for segregation of waste and facilitate collection of segregated waste. Additionally, they transport and compost wet waste, recycle dry waste and sell organic manure. The Swachha Sathis are responsible for creating awareness of segregation and monitoring of the collection of waste; they lend support in weighing the waste generated in various households to help prepare city-specific action plans based on dry and wet waste generated. Each Swachha Sathi is typically responsible for engaging with 600 households and one supervisor for every 2400 households. They are paid a monthly honorarium by the state. Currently, about 185 micro-composting centres/material recovery facilities are run by SHGs.

Another advancement of participation of community groups came when the drink from tap mission was adopted working with SHGs as a core strategy. The Public Health Engineering Organisation is the agency for providing water supply to the cities, but with universal water supply being the aim, the workload increased exponentially. Therefore, SHGs were brought into the programme. Initially they were involved through the formation of a water and sanitation committee, which was then modified to the concept of water monitor and further modified to Jal Sathis. The Jal Sathis, who are members of Mission Shakti SHGs, act as a link between public and water supply authorities and perform a number of functions (including monitoring the water quality in the taps after being trained and equipped to undertake water quality indicator tests) on a performance-linked incentive basis. The Jal Sathi model demonstrates that urban decentralisation and partnership with community-based institutions is key to improved and responsive governance.

A similar concept has been applied to SDAs, which unlike SHGs are tied to the location of their slum, under the Jaga Mission in 2018. SDAs have been created in every slum with a minimum 50% women members. They assist in household surveys and mapping of slums, ensure participation of slum dwellers and undertake community mobilisation. As the mission progressed from granting of land rights to upgrading of slums, to upgrading infrastructure and maintain the same, roles and responsibilities of the SDAs increased substantially. SDAs as implementing partners (IPs) in slum upgradation work are empowered by the urban wage employment initiative (UWEI) programme converged under the Jaga mission. The MUKTA scheme learning from the UWEI experience in converging with the Jaga mission from initiation aims to address two important objectives: community asset creation, operation and maintenance by identifying and implementing labour-intensive projects in the cities, and financially strengthening the existing ecosystem of CBOs such as Mission Shakti SHGs and SDAs by positioning them as IPs in executing these projects.

Last but not least, a new scheme, 'Ama Pokhari' has been launched to rejuvenate urban water bodies naturally and sustainably with the help of Mission Shakti groups. This initiative highlights the potential within communities, especially women, to lead this transformation. Women-led Mission Shakti SHGs are central to the efforts, actively engaging in de-weeding, desilting and plantation activities while serving as guardians of the revitalised water bodies. Initially, the programme faced scepticism: 'How can housewives in SHGs understand waterbody rejuvenation?' 'How can embankments survive without heavy structures?' Despite doubts, Ama Pokhari has become a model of social empowerment and environmental restoration. It equips women with skills and resources to shape their city's future and showcases their leadership. For the first time, fresh graduates from local colleges have been given the chance to supervise the rejuvenation process as interns. Their involvement has helped ULBs scale up activities and achieve high targets.

13.4 COMMUNITY PARTICIPATION CENTRED ON PROMOTING RESILIENCE OF THE URBAN POOR

Although economic prosperity is often seen as a corollary of urbanisation, another – oft-neglected – corollary is inequality, poverty and vulnerability within cities. Although cities are often the focal points of economic prosperity and activities that attract investments to improve the overall development of the country, they are also hubs of poverty, deprivation and vulnerability. Rising inequality, poverty and social exclusion can thus derail development and constrain the economic gains of urbanisation. Urban residents are often forced to reside in slums and informal settlements, with minimal or no access to basic services. Although the need to create more inclusive cities has been recognised nationally and internationally, responsive strategies for this are limited. Building inclusive and resilient cities remains a challenge that governments at all levels are dealing with, and the Government of Odisha is no exception.

Overall urban inclusion can only be achieved when all urban residents, especially the urban poor and vulnerable have opportunities to reap the benefits of economic growth. The government recognised that high poverty and increasing marginalisation of livelihood have forced many people to migrate to other states, thus making Odisha one of the foremost migrant-sending states in India. This was also highlighted during the COVID-19 pandemic, as many migrant workers started returning to their towns and villages in Odisha owing to the nationwide lockdown. Recognising the interlinkages between poverty and urbanisation, the Government of Odisha's urban reforms have been driven by the primary aim of poverty reduction, provision of sustainable livelihood opportunities and empowerment of the urban poor.

Community engagement has been one of the key features of the urban development programmes in Odisha to ensure socio-economic inclusivity in cities, and this has built community and city resilience. Figure 13.1 explains how each pro-poor programme of the government is contributing to building social, economic and physical reliance in cities and communities.

Slum upgrading has provided the urban poor with access to basic infrastructure, such as in-house piped water supply, electricity and toilet among others, such interventions not only ensure resilience – through improved public health outcomes – for the slum dwellers but also for the city at large. Detailed project reports prepared for each slum, for the identified infrastructure gaps, focus on climate-sensitive design and local material usage such as the use of paver blocks for roads, light-emitting diode (LED) bulbs for streetlight, construction of toilets with a septic tank and so on. These technologies are expected to result in subsoil percolation and groundwater recharge, energy

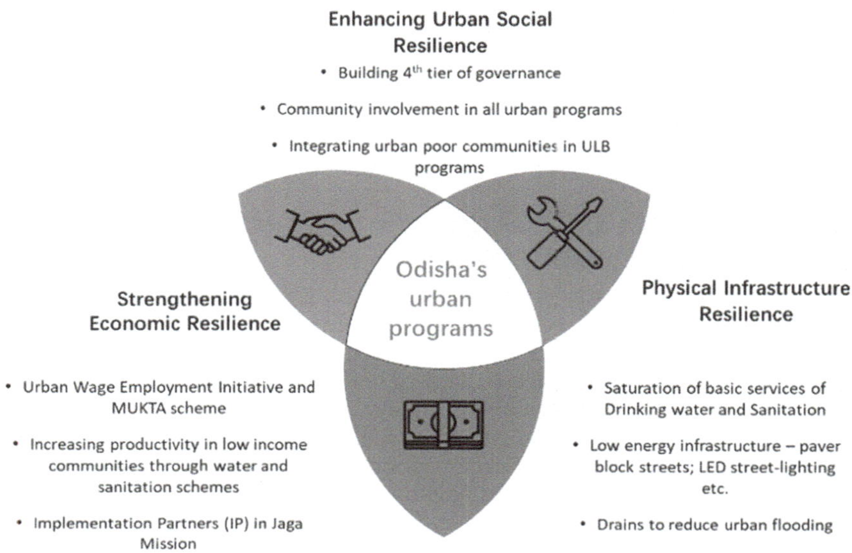

Figure 13.1 Building resilience through Odisha's urban programme. *Source*: Authors' analysis.

conservation, fuel conservation, reduction in carbon footprint, groundwater contamination, water body contamination and health hazards, signalling a positive environmental impact.

In the wake of the COVID crisis, Odisha government introduced the urban wage employment initiative to employ the urban poor by engaging them in the creation of urban infrastructure. Such an initiative not only increases their income, but also indirectly facilitates them to invest in education, health, better housing – all of which contribute to human development while making a long-term impact on their lives. Infrastructure creation is undertaken by the slum community themselves through SDAs under the UWEI scheme; this creates livelihood opportunities for the poor and ensures community ownership of the assets created for slum upgradation, fostering community and economic resilience.

13.5 UPHOLDING DIGNITY IN ADDITION TO ECONOMIC PROSPERITY

The term 'dignity', while often mentioned across a plethora of developmental documents, is seldom contemplated in depth. 'Amartya Sen (and others) transformed the development lexicon by defining development as freedom rather than just economic or even social progress, and the concept of dignity takes us a step further along that road' (Glennie, 2015).[2] Although poverty and inequality are conditions at the heart of the urban challenges, the lack of

[2] https://www.theguardian.com/global-development/2015/jan/28/dignity-sustainable-development-goals.

dignity among those at the bottom of the economic ladder is another challenge in that path towards eradicating poverty and inclusive development and must be addressed in urban policy.

Inclusive urban growth cannot lose sight of the need for a dignified urban life, and this has been foremost in Government of Odisha's mind, as it has implemented most of its urban development programmes as discussed in various chapters earlier. A prominent example is the Garima scheme for those involved in hazardous sanitation work or manual scavenging. As already discussed, these workers have long been subjected to untouchability and stigmatisation owing social and economic isolation due to exposure to 'polluting' faecal matter as part of their work. The Garima scheme was launched specifically to address the safety, dignity and socio-economic issues confronted by sanitation workers. In another example, the most important principle of the state's Aahaar scheme is that an intervention for the poor need not be of poor quality. Thus, the superior quality infrastructure created along with enforcing of safe hygiene practices across the entire value chain not only ensures the provision of safe and nutritious food to the urban poor, but also upholds dignity of the urban poor availing this scheme. Each of the pro-poor HUDD initiatives is designed keeping the dignity of the target beneficiaries in mind.

13.6 BUILDING THE FOURTH TIER OF GOVERNANCE

The Odisha approach to urban development is a unique model – fostering the fourth tier of governance, as decentralisation is conceptualised, designed and executed as an integral part of programmes/schemes. The success that the HUDD has experienced in deploying community participation in all government schemes has contributed to conceiving and craft area-based community groups, that is, the SDA as the fourth tier of governance in cities. Involving citizens and communities as vehicles to help achieve objectives of HUDD missions has been a cross-cutting approach across schemes and has ensured that employment for the urban poor is generated. Mission Shakti groups, Swachha Sathis, Jal Sathis, SDAs and so on have been acting as active IPs of ULBs in Odisha and have been critical in delivering the objectives of each scheme/mission of the state government. Building on this positive experience the Government of Odisha has been working systematically towards building the capacities, institutional strengthening and economic empowerment of the SDA as the fourth tier of governance.

Some key enabling actions by the state government through its schemes discussed in the earlier chapters towards nurturing community participation towards building the fourth governance tier are:

- Partnership for service delivery in water, sanitation, the subsidised food programme, employment schemes and other municipal functions.
- Developing business models, SOPs, guidelines, protocols, standardisation, model service contracts to engage CBOs.
- Systematic capacity building of CBOs and other stakeholders.
- Community engagement through CBOs as not only a means but an end in itself – not only a process issue but in most cases an essential purpose of the scheme itself.

- Supporting the urban poor in progressing towards non-conventional, skilled tasks with managerial and technical capacities with every new intervention.
- Focus not just on financial benefits but on overall growth and professional competencies of CBOs, leading to a sense of pride, accomplishment and dignity.
- Building cooperation by sensitisation and orientation of various government functionaries to partner and work with CBOs.

The success with operationalisation of community participation in all schemes gave the Government of Odisha confidence to conceive and build the area-based CBO, that of the SDA as a fourth tier of governance. This concept of the fourth tier, under the ULB which is enshrined in the Constitution of India as the third tier, furthers the spirit of decentralisation that the 74 CAA had set off.

With this in view, on 10 November 2022, the Government of Odisha announced the recognition of SDAs as the fourth tier of government, making them partners in the development and governance process. Municipal bodies have signed memorandum of understanding (MoUs) to empower SDAs to self-govern the Biju Adarsh Colonies. Additionally, 25% of the municipal fund has been statutorily earmarked for SDAs.

With the rapid implementation of the slum upgrading and land rights programme under the Jaga mission, as explained in the earlier chapters, the slums are being transformed into Biju Adarsh Colonies. The HUDD of the Government of Odisha has ensured that all ULBs have signed MoUs with SDAs as they emerge as Biju Adarsh Colonies with all basic services, land rights, housing and get integrated into the city. The MoUs involve the ULB giving the rights of maintenance of the community infrastructure to the SDAs. The ULBs are also empowering the SDAs to collect user charges on their behalf and provide them with grants where required. MoUs have been signed by each of the 115 ULBs with the SDAs in their jurisdiction.

Furthermore, the ULBs are now statutory directed under the Municipal Act to devolve 25% of the municipal budget to SDAs and resident welfare associations of the Biju Adarsh Colonies. These funds could be used as per proposals by SDAs to develop local community development programmes across, infrastructure, services or socio-economic development programmes. The HUDD and ULBs are also ensuring that regular accounts are maintained by SDAs for which a regulatory structure and guidelines and capacity building programmes are being held. With this Odisha has demonstrated how a fourth tier of government, which is closer and more responsive to the needs of the urban poor can be built. This effort has rapidly evolved since 2018, when community participation was introduced in urban services in a structured manner at scale.

The fourth tier of government is expected to further transform the lives of citizens across ULBs in the state and truly reflects the empowerment of communities alongside decentralisation in urban governance – a model unique in India, and one if emulated in other geographies and jurisdictions could benefit all.

doi: 10.2166/9781789064926_0215

Chapter 14

Key facets of the Odisha urban development model

ABSTRACT

This chapter discusses the state's overall strategy for ensuring inclusive growth and socio-economic empowerment of all citizens, through partnerships, collaborations and joint initiatives of the government, civil society organisations, academia and corporates: piloting before scaling. It further summarises the learnings from the state's experience of implementing various urban development programmes as discussed in different chapters of the book, and how other Indian states can adopt these learnings for their respective urban development programmes.

Keywords: Inclusive, sustainable, phased, partnership, inter-sectoral financing

14.1 AN URBAN DEVELOPMENT MODEL ALIGNED WITH INTERNATIONAL TARGETS

Over the last few years, in tandem with the increasing urban trajectory of Odisha, the state government has bolstered mechanisms to ensure urban infrastructure development, quality service delivery, decentralised governance and community partnerships, keeping the people at the centre with special focus on the urban poor. Even as the state has dealt with a plethora of disasters over the last few years, and now the COVID-19 pandemic, its urban governance strategies have fostered resilience and sustainability in the cities, along with promoting the welfare of the people, resulting in achievement of the sustainable development goals, especially SDG 11 (Sustainable cities and communities).

The state's model of urban development is also in alignment with the UN-HABITAT's Global Campaign on Urban Governance, launched in 2000 with the aims of enhancing the capacity of local governments and other stakeholders to practice good urban governance and to raise awareness towards making urban development more sustainable and inclusive.[1] In addition, the initiatives of the Government of Odisha to tackle urban issues spur the state's movement towards the achievement of the New Urban Agenda (NUA). The NUA was unanimously adopted at the United Nations Conference on Housing and Sustainable Urban Development (HABITAT III) 2016, ushering a paradigm shift in city planning, building and management.

In today's rapidly urbanising world, the challenges of governing cities have become more complex and intricate. The NUA recognises the difficulties that arise from multilevel and multi-stakeholder governance systems that are based on interactions between different levels of government, citizens and a wide range of non-governmental actors, including the private sector. To address these challenges, the NUA recommends that urban governance responds clearly to the increasing demands of citizens, and thereby addresses the rights to the city, by creating enabling legal and institutional frameworks at different governmental levels. This approach ensures that the more marginalised and disadvantaged sections of society are prioritised in each intervention. Therefore, the right to the city is realised through a cross-cutting policy approach.

The state of Odisha in India has been at the forefront of implementing such an approach. The Government of Odisha has been working on various urban development models in the state, taking into consideration all 17 SDGs outlined by the United Nations. The Jaga mission, launched in 2018, aimed to provide land rights to slum dwellers and improve their living conditions. The initiative was aimed at achieving SDG 11, which aimed to make cities and human settlements inclusive, safe, resilient and sustainable. The government also launched the urban wage employment initiative, aimed at providing employment opportunities to the urban poor, thereby promoting SDG 1, which aims to eradicate poverty.

The faecal sludge and septage management (FSSM) project was launched to promote clean and healthy living conditions and to achieve SDG 6, aimed to ensure the availability and sustainable management of water and sanitation for all. Additionally, the drink from tap initiative, launched in 2020, aimed to provide safe and clean drinking water to all households in the state, contributing to SDG 6. The Aahaar scheme was launched to provide affordable meals to urban poor, promoting SDG 2, which envisioned to end hunger.

The government also launched the Garima scheme, aimed at providing skill development and employment opportunities to the urban poor, contributing to SDG 8, which focuses on promoting decent work and economic growth.

Overall, the government of Odisha is striving to achieve all the SDGs through its various urban development models, promoting a sustainable and inclusive future for its citizens. The government's efforts towards sustainable urban development models are not only contributing to achieving the SDGs but also setting an example for other states and countries to follow.

---

[1] https://www.ucl.ac.uk/dpu-projects/drivers_urb_change/urb_governance/pdf_trans_corrupt/HABITAT_Global_Campaign_Good_Urban_Governance.pdf.

Busting myths

The philosophy in governance emphasises that interventions for the poor should not be substandard. Demonstrating this philosophy, Aahaar centres, which provide affordable meals for the urban poor, are equipped with high-quality infrastructure, amenities and services. City governments are advised to construct and maintain public and community toilets to quality standards comparable to those in airports and shopping malls, with requisite funds provided.

Adequately funded and well-executed interventions not only impact the lives of people lacking sufficient access to services but also influence the conservative mindset and perception of decision-makers at the implementation level. Some initiatives have challenged prevailing myths and beliefs about the urban poor.

For example, the Aahaar programme counters the notion that providing subsidised food encourages laziness among the poor. Empirical data show that only 45 000 to 50 000 meals are consumed daily from 167 Aahaar centres across the state, against a sanctioned capacity of 0.1 million meals. This indicates that providing subsidised food does not lead to dependency on these centres.

Another myth dispelled by an initiative is through the Jaga mission. When the Odisha Land Rights to Slum Dwellers Act was passed in 2017, concerns were raised that granting land rights would encourage large-scale encroachment and migration of the rural poor to urban areas. However, over the past 5 years, no city in Odisha has experienced abnormal growth of slums, confirming that people do not leave their villages and livelihoods merely because the government provides non-transferable rights to 300 square feet of housing space.

– G. Mathi Vathanan

The urban governance model of the Government of Odisha presents a paradigm shift towards a more decentralised and inclusive urban governance approach. This chapter summarises the learnings from the key facets of the initiatives of the state government for replication – to enable city and state governments across the country and the world to strengthen their urban planning and governance regimes.

14.2 STRATEGIC APPROACH

Urban planning can no longer be seen as a unidimensional, static and technical process. Rather, the dynamism of cities has compelled urban planners to address the challenges of urbanisation through a multi-sectoral approach, converging various perspectives. As cities grow rapidly and irrevocably, optimal management of city resources and capacity building through strategic planning approaches has become indispensable.

The Government of Odisha has constantly approached urban planning and development through multi-pronged strategies that help the city administrations to respond to rapidly changing situations as they endeavour to improve the quality of life of urban inhabitants. Although strategic planning processes are in no way a substitute for spatial planning, they guide the development process in the direction of those interventions identified by all stakeholders as ones that would improve the overall quality of life in the cities, even against the background of continuous changes in the city. Owing to the dynamism of urban areas, the state government has also attempted to implement its initiatives, schemes and policies through an iterative and consultative process, which provides scope for constant learning and fine-tuning of the proposed interventions, before arriving at the final set of decisions.

The strategic approach to urban development adopted by the Government of Odisha primarily encompasses two methodologies of intervention: research and development of laws, policies, regulations, institutions and financing schemes; and implementation through piloting and scaling up.

14.2.1 Research and development of laws, policies, regulations, institutions and financing schemes

(i) ***Deep understanding of the local context in terms of politics, administration, people's requirements and social norms***: Urban governance has to be configured according to the context in which it operates. Often, the state-level laws, policies, regulations, institutions and financing mechanisms are not adequately adapted to the local contexts of cities. In such a scenario, there is often a disconnect between legal, administrative and fiscal frameworks, lack of clarity in the devolution of responsibilities between different levels of governance and introduction of regulations that are often contradictory. In addition, these frameworks are often too rigid to respond to the rapidly changing dynamics of urbanisation. To ensure cooperation at the organisational, local and regional levels, there is a need to understand the various local settings that can support or hinder the implementation and outcome of an urban governance intervention.

The Government of Odisha has consistently adopted strategies that are based on a thorough understanding of the local contexts - economic, political and social - including an understanding of the formal and informal processes and structures. In accordance with this consultative and inclusive strategy, the state government partners with local NGOs, key community influencers, slum dwellers' associations (SDAs), self-help groups (SHGs), among others, to attain a ground-level understanding of the nuances of the urban ecosystems. Additionally, the state government also recognises the need to look beyond the national-level data sets, which are often outdated and are limited in their ability to capture local intricacies. Consequently, the formulation of the laws, regulations, policies, institutions and financing schemes is preceded by household surveys, spatial surveys, drone mapping exercises and key informant interviews.

For example, the formulation of the Odisha Septage Management Guidelines, 2015, Odisha Urban Sanitation Policy, 2017 and Odisha Urban Sanitation Strategy, 2017 was based on consultative processes between the urban local body (ULB) representatives, cesspool operators, treatment facility operators and the owners of premises. These strategic approaches were crucial in addressing the regulatory gaps in FSSM at the ULB level and enabling equitable and sustainable access and use of sanitation services.

(ii) ***Developing state-specific policies for each sector, leveraging Government of India's schemes***: Policy frameworks at the national level – including the provision of fiscal support to state governments – can guide local institutions and actors towards effective urban governance. This also involves the articulation of an urban vision that outlines the arrangement of resources and institutions. State- and city-level frameworks can only be effective if the national policies foster a favourable and flexible environment.

In the case of urban policymaking, national policies often focus on cities as a 'whole', instead of viewing them as individual, economically vibrant units. Furthermore, as cities are constantly expanding in India, designing urban governance frameworks that cater to the dynamism of urban centres across the country is next to impossible for the national government. Thus, the national government allocates the responsibility to the state governments to design appropriate urban governance regimes and to create accountability mechanisms for the implementation of policies formulated at the state level.

In most scenarios, there is no need to reinvent the wheel. That said, the Government of Odisha has attempted to leverage the national-level policies to formulate state-level policies that are customised to suit the socio-economic outlook of the state and encompass a host of forces such as labour markets, households, land use and public service delivery. In the face of evident resource constraints in the state, such state-level policies also enable a targeted dissemination of the benefits to the urban poor and the most vulnerable groups in the city.

For instance, furthering the objectives highlighted in the National Faecal Sludge and Septage Management Policy, 2017, the Government of Odisha outlined its own set of policies and guidelines in the form of the Model Faecal Sludge and Septage Management Regulations in 2018. Applicable across the entire sanitation value chain, they outline the roles and responsibilities of the stakeholders. Similarly, the state government introduced the landmark legislation – the Odisha Land Rights to Slum Dwellers Act – to enable a rapid uptake of the BLC subsidy under PMAY-U.

14.2.2 Implementation through a process of sub-pilots, pilots and scaling up

Across its schemes, the state government has used specific (i) sub-pilots, (ii) pilots and (iii) scaling-up processes to ensure their smooth implementation. As discussed in each of the programme chapters, this overriding process of increasing scales is a standard practice that has been adopted across all infrastructure programmes of Odisha's nodal urban development agency, the Department of Housing and Urban Development (HUDD), since 2015. Although this approach can be seen from the earliest period over the last 8 years, these processes have been streamlined and are now well embedded.

This approach simultaneously addresses several issues that are faced in taking up innovative ideas within government. The first issue that a step-by-step process like this address is the element of administrative and financial risk that government officers can take. The second key element is that the technical viability of an idea can be tested for the Odisha context and lessons to be kept in mind before scaling up can be incorporated. The sub-pilot and piloting phases are conducted with the participation of the senior-most state government officials, key district- and local-level officers and particular technical partners. The third advantage of such an approach is that the capacities of a core set of actors in government and in supporting organisations is built who then act as mentors/guides during the scale-up process. Finally, this process broken up in three phases ending up in scaling-up services across all cities allows the key decision makers to be self-critical and continuously learn lessons while adopting a flexible, pragmatic approach, instead of a static bureaucratical one, to implementation (Figure 14.1).

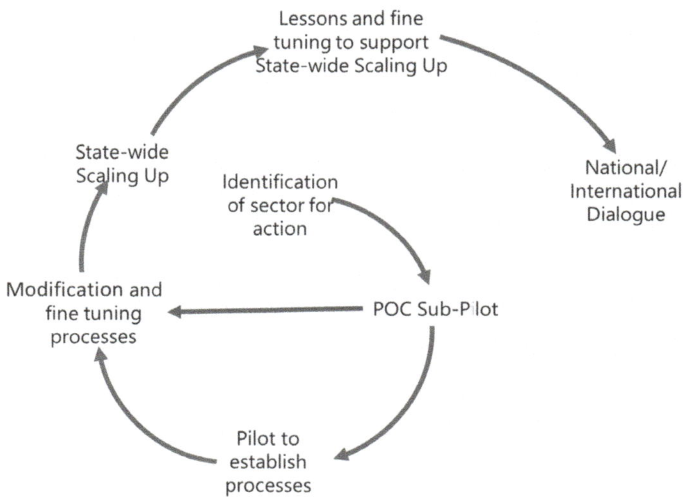

Figure 14.1 Theory of change. *Source*: Authors' analysis.

Sub-pilot phase: The sub-piloting is a hands-on project development exercise carried out in a limited geographical area in the field to test the viability of the primary idea of the scheme. Sub-piloting a project is necessary to evaluate whether a programme or intervention will work in the local context. Owing to the multi-sectoral approach adopted in most developmental projects and the concomitant interface between several stakeholders including ULBs, local communities and civil societies, such a feasibility study to assess the viability of implementation and scaling up is crucial. In these circumstances, sub-piloting is therefore applied to gain lessons on viability as well as key components of the project that would need special attention.

The 'drink from tap' initiative, which ensures 24 × 7 drinkable tap water, is one example of a sub-pilot. The project was initially tried in a small settlement called Ishaneswar Basti in Bhubaneswar with a population of 2000 people so as to control for the risk. Given that drinking water supply is a particularly sensitive service, the risks associated with it are high. An example of a risk faced by the sector is that if water supply is disrupted, it will cause huge inconvenience to the community and could lead to political unrest. Other than this, if the quality of the tap water is not suitable for drinking, it could also lead to disease in the community and disrupt social harmony. As water is a sensitive matter with high associated risks of contamination, extensive field experimentation was needed; it was necessary to conduct this experiment in a smaller geographic area under controlled conditions to minimise the risk while ensuring an adequate back-up response system. Hence, the sub-piloting covering 2000 people was designed with the understanding that if any major issues came up, the government would provide immediate assistance through tanker supply.

Pilot phase: The sub-piloting offers limited learnings on the process required for full-on scaling up and, therefore, it is followed up by an intermediate piloting phase. Piloting is a small-scale implementation programme in which the suitability of a novel strategy is examined before it is integrated into the current operations. It is essentially aimed at deciding the modalities of the implementation of the programme that would be easy to implement at scale and provide quickest results in a sustainable way. Implementing a pilot before undertaking a project at its stipulated scale is an innovation in programme implementation that has led to successful initiatives.

In the case of 'drink from tap' 24 × 7 water supply, after the sub-pilot proved the viability of such a programme, the HUDD started a pilot exercise in 12 zones in two cities covering a population of 0.18 million. On the other hand, in the case of the Jaga mission a sub-pilot was not undertaken; it went directly into a piloting exercise. The initial step was to conduct a pilot exercise in 50 slums across two districts and nine cities. The pilot primarily served to maximise the learnings and identify the do's and don'ts for the full-scale implementation of the land rights to slum dwellers programme. This led to the creation of field-tested standard operating procedures (SOPs) and established a step-by-step process for state-wide implementation.

Following the guiding principle of 'leaving no one behind', all the flagship programmes of the Odisha government are created to be implemented across all cities in the state. With the state facing severe capacity constraints at the city level about technical, financial and managerial factors, the piloting strategy has been able to develop a well-thought-out and validated implementation template with detailed SOPs for implementation at the city level. Thus, piloting is a prerequisite: it enables flexibility of reassessment and easy amendments in the designed interventions in a low-risk manner, while also aligning the institutional mechanisms necessary for scaling up the project.

***Scaling-up phase*:** With the success of a project in its pilot phase, the innovations are expanded and replicated to reach more people and to broaden the scope and effectiveness of an intervention by scaling up. Attention to scaling up necessitates dedicated and systematic planning, supported by a clear institutional set-up, to implement the innovations on a larger scale and achieve broader impact. As a part of the scaling-up process of innovations, the Government of Odisha has attempted to create conducive conditions that allow individuals, administrations and teams to adapt a solution to their unique circumstances and make it their own.

The Government of Odisha has rapidly scaled up the pilot projects that have demonstrated a dramatic impact trajectory. For instance, the pilot projects under the 'Sujal' or drink from tap mission were implemented in five water supply zones in Bhubaneshwar and two in Puri. Soon, the pilot initiative in Puri was scaled up to cover the entire city, making it the first city in India to achieve drink-from-tap quality water supply with 100% metered connections.

It must be noted that, typically, innovations are tested in pilot projects with special organisational, financial and human resources, which will not be available at scale. Under such circumstances, scaling up must be accompanied by simultaneously balancing between desired outcomes and practical realities and constraints. Furthermore, although the initial plan provides the foundation, strategic thinking during the scaling-up process needs to be dynamic to adapt to the multiple factors that affect the expansion and should allow adjustments to the strategy wherever necessary.

14.3 PRIMARY FACTORS BEHIND SUCCESSFUL IMPLEMENTATION OF ODISHA'S URBAN INITIATIVES

The success of the initiatives of the Government of Odisha can be attributed to risk-taking ability through pilots and innovations; rapid scaling up of successful pilots with cross-department support and courage of conviction demonstrated through new laws, policies and regulations, based on success in pilot implementation.

14.3.1 Risk-taking ability

The Government of Odisha is committed to implementing urban development interventions that can be applied across the state. For optimal interventions against a background of resource constraints, the state government's first step often entails taking the risk of piloting the projects in selected regions. The qualitative and quantitative information derived from the pilot project is statistically and spatially analysed and the results are interpreted. Based on the success or failure of the pilot programme, a decision is made to improve the programme design through a consultative process, if required, and scale up the project.

14.3.2 Rapid scaling up with cross-department support

To achieve impact at scale and broaden the scope and reach of innovative pilot programmes, the Government of Odisha aims to scale up and replicate the pilot programmes. The rapid and effective scaling up of pilots is predicated on a clear institutional set-up, synergetic functioning of and between government departments and sustained cross-department support. Moreover, attention to scaling up necessitates dedicated and systematic planning to implement the innovations on a larger scale and achieve broader impact.

Successful pilot implementation in Puri resulted in a city-wide scaling up of the project, consequently making it the first city in India to achieve 100% metered connections and drink from tap water quality. Another example of successful piloting emerges from the slum delisting and upgradation component of Phase II of the Jaga mission, wherein, surpassing the initial target of 500 slums, 550 slums across 28 ULBs were transformed into 'Adarsh colonies' within just 6 months of rolling-out of the SOP.

Changes that nurture growth

The approach to implementation adopted across programmes by the HUDD can be said to be key to the success of each programme. The approach of incremental risk taking and learning through the sub-pilot, pilot and scaling-up phases allows for building capacity among state- and local-level government functionaries while also building knowledge on each of the sector interventions. By adopting this approach in each sequential scheme, the Government of Odisha has increased its state capacities and HUDD has emerged as a learning organisation over the last 8 years. Alongside, the urban sector in Odisha has become a site for innovation at scale.

This has continuously attracted numerous development agencies and sector professionals to work on programmes in Odisha. The symbiotic, positive and reinforcing knowledge environment these processes have created has not only resulted in the numerous awards that HUDD has received in the last few years but is reflected in the improved situation of lower income communities across cities and has surely contributed to the economic gains of the state over the last few years.

More research needs to be undertaken in Odisha on the interconnections involved in changes to internal systems in a department's functioning that contribute to creating an enabling environment for positive change on the ground, stimulating socio-economic development. Hopefully, this first effort in that direction will trigger more such work.

– Shubhagato Dasgupta

14.3.3 Courage of conviction demonstrated through new laws, policies and regulations

Changes in one element have repercussions for the others. Dynamism is vital in strategic urban planning which shows governments courage of conviction. Any approach to urban planning and development needs to recognise the intricate networks of interactions and influences between different stakeholders, which should be accounted for to ensure successful innovations and interventions. In addition to encouraging collaborations and behaviour change, implementation of an innovative scheme often requires the creation of an enabling environment through new laws, policies and regulations.

Changes in one element have repercussions for the others. Dynamism is vital in strategic urban planning. Any approach to urban planning and development needs to recognise the intricate networks of interactions and influences between different stakeholders, which should be accounted for to ensure successful innovations and interventions.

Towards the achievement of the national and international goals of sustainable, inclusive and resilient development, the Government of Odisha has constantly endeavoured to put in place an iterative learning mechanism for its schemes, policies, laws and regulations. It has ensured that programme implementation is constantly fine-tuned by learning from ground-level realities and relationships, without losing sight of the overall objectives and outcomes of the interventions. For instance, Government of Odisha notified the first of its kind state-level policy on sanitation, namely Odisha Urban Sanitation Policy (2017), which spotlights the need for FSSM in urban areas – responding to a high prevalence of on-site sanitation in the urban areas.

14.4 CITY-WIDE UNIVERSAL INFRASTRUCTURE PLANNING

City governments across the world face major challenges related to the environment, economy, sustainability, social inequality and lack of universal service delivery. The variety of problems associated with urban growth jeopardise the sustainability of cities, and obstruct the stride towards the achievement of SDGs and NUA. However, most of the corrective measures that are implemented in India today are defined over specific coverage domains, without much scope for universal coverage. Sustainable urbanism will require a universal approach to city management and urban planning that aims for long-term viability through a reduction of operational and implementational costs and improving social equity and well-being.

Each urban initiative of the Government of Odisha is designed for universal coverage, wherein even the poorest parts of the city are provided services to bring them one step closer to being integrated into the overall city fabric. Adhering to the principle of 'leaving no one behind', the schemes are designed for a saturation-based approach, benefitting all citizens in an equitable manner.

For instance, even as the second phase of the Jaga mission prioritised slum upgradation and delisting to bring the tenable slums on par with the rest of the city, there remained households at the city margins that occupied environmentally hazardous land or lived under a significant threat of eviction.

In line with the government's firm resolve to integrate all slum dwellers into the city fabric, it has implemented a range of measures such as land re-classification, land exchange, slum re-densification and so on to ensure as much in-situ settlement of land as possible, subject to written consensus of the slum dwellers through the SDAs. Hence, in the second phase of the mission, the government is upholding the citizens' 'right to the city' by undertaking processes to address the untenability aspect and provide them with services and housing.

14.5 INNOVATIVE FINANCING APPROACH

The 74th Amendment of the Constitution recognised the ULBs as the third tier of the government and assigned several municipal functions to the municipal governments. However, this increase in responsibilities was not matched with adequate financial empowerment of the ULBs, thus restraining the capacity of the cities to act as 'engines of growth'. Municipal revenues/expenditures in India have been stagnating at around 1% of the GDP for over a decade.[2] Among the numerous challenges to be tackled in urban development, instituting appropriate financing mechanisms to undertake development reforms is a key element that needs to be reflected in national and local authority budgets.

The state of Odisha was no exception when it came to the financial capacity of the ULBs to implement urban reforms. However, despite these challenges, the Government of Odisha rose to the occasion by implementing innovative financing mechanisms, such as developing interlinkages between different national- and state-level schemes to leverage their funds and piloting new innovations with the existing state budgets.

14.5.1 Interlinkages between schemes for funding through a bottom-up approach

To successfully address urban issues and strengthen urban economy and livelihoods, there is a need to optimise efforts through inter-sectoral linkages. Convergence between various programmes brings in synergies between different government programmes vis-à-vis funding, planning and implementation, and ultimately advances the movement towards sustainable development.

The Government of Odisha has facilitated the convergence of schemes and programmes, which has also contributed towards an increase in economic opportunities, income generation and overall poverty alleviation. Furthermore, such convergence efforts have led to a greater sense of participation, partnership and accountability at the grassroots level, thereby enhancing the quality of outputs and the social capital in the state.

In Odisha, to manage safe and sustainable FSSM, existing programmes and resources were utilised for sustaining FSSM-related activities; for example, Atal Mission for Rejuvenation and Urban Transformation (AMRUT) funds were used to provide capex for sewage treatment plants. Furthermore, service provisioning,

---

[2] https://fincomindia.nic.in/writereaddata/html_en_files/fincom15/StudyReports/State of Municipal Finances in India.pdf.

capacity building and awareness programmes were facilitated by the National Urban Livelihoods Mission and National Health Mission (NHM) while Smart City programme and ULB funds were utilised for single-window system and information and communications technology for cesspool requisition.

The implementation of the 'Universal Access to Piped Water Supply Programme' required a huge input of financial resources. Owing to the scale of this ambitious programme and the limited availability of financial resources, a strategy was developed to leverage funds from a variety of sources such as AMRUT and Urban Infrastructure Development Scheme for Small and Medium Towns Scheme, State Plan funds and other sources such as the Development Management Fund and Odisha Mineral Bearing Areas Development Corporation. In addition, funds were also sourced from Odisha Urban Infrastructure Development Fund and corporate social responsibility funds from the National Thermal Power Corporation.

14.5.2 Piloting new innovations within existing budgets

The Government of Odisha has demonstrated a long-standing commitment to welfare and inclusive development, with ongoing schemes that have put that commitment into practice. Because of resource constraints, especially amid the COVID-19 pandemic, the government has attempted to converge financial flows across sector-wide schemes by combining the implementation configurations to suit the activities.

The flagship mission of the state – the Jaga mission – is being implemented by utilising the existing human resources, as its establishment cost has been already sourced from the annual state budget, with no additional budget required. The mission illustrates various successful innovations in converging resources from various other initiatives. For instance, the programme converged with Mission Basudha, which has a dedicated fund of Rs 2360 million (USD 31 million) to enable drinking water supply while upgrading slums, rather than creating a separate component for just water supply. Furthermore, community-based organisations such as Mission Shakti, women's SHGs and their area-/city-level federations and SDAs engaged as implementing partners in construction are funded under the Urban Wage Employment Initiative (UWEI) scheme. UWEI is a component under UNNATI, a programme included in the state budget to fill critical gaps in urban infrastructure. The UNNATI initiative is also bolstered by an integrated funding mechanism, wherein more than 500 projects have been given financial approval directly at the state level, thereby reducing delays and leading to a more seamless implementation process.

14.6 DATA-DRIVEN URBAN MANAGEMENT ENSURING TRANSPARENCY IN GOVERNANCE

The digital era has dramatically altered the way our societies operate. New technological developments have ushered in unforeseen opportunities for businesses, citizens and public agencies. Yet, public authorities often find it difficult to respond to these new developments. Many city governments face severe capacity constraints and lack the vision for sustainable urban growth owing to a lack of updated data on key attributes of the cities, such as population, economy, informality, vulnerability and so on. Building capacity in urban planning, budgeting and public asset management can be substantially bolstered through the adoption of digital governance mechanisms and data gathering.

For enhanced data-driven decision-making in Odisha, the government has implemented structural, organisational and procedural provisions as well as overall governance systems that have ensured transparent, accountable decision-making in the urban interventions of the state.

14.6.1 Improving implementation and operation and maintenance through digital management systems

Lack of adequate data restricts urban situation analysis and undermines planning and management of service delivery. Many state- and local-level governments face a persistent lack of access to localised data which impedes their ability to take informed decisions and better prioritise local policies. It is imperative that data collection is integrated within the urban governance system, in a way that also recognises civil society organisation, guarantees and promotes participation of citizens and allows access to public data to promote transparency and accountability in public institutions.

The urban interventions of the Government of Odisha recognise that monitoring and evaluation are the cornerstones of any successful intervention. The government's data-driven approach towards urban governance is evident in its interventions in solid waste management (SWM), FSSM, drinking water and the Jaga mission, among others.

Adopting a digital route to monitoring SWM in the state, the Ama Sahara app was launched on 31 August 2020. The app fulfils the dual goals of operational management and monitoring by enabling the collection of data of the day-to-day haulage of solid waste, the total quantity of waste processed and its disposal, and the revenue generated from the sales of compost and recyclable non-biodegradable waste. It also allows citizens to make payments for SWM services online and request garbage pick-up or orientation on segregation, while also giving them the ability to state their grievances, thus ensuring transparent and rapid service delivery.

Similarly, in the case of water supply, to integrate the data from assets set up under different projects at different periods of time, data on existing water supply assets were collected from a variety of primary and secondary sources including previous project reports, consultations with field officials, group discussions with concerned staff, interactions with local communities/ beneficiaries and so on. Subsequently, the data thus collected were mapped over a geographical information system (GIS) platform to ensure easy access and to facilitate decision-making. Furthermore, through automated systems such as programmable logic controllers/supervisory control and data acquisition installed all along the value chain, real-time water data are collected and displayed through digital display boards at key public locations. This also helps reduce manpower requirements and facilitates decision-making for uninterrupted and reliable water supply.

The Government of Odisha's Jaga mission is another striking example of data-driven implementation and decision-making, across various points of the value chain, for successful uptake and scale-up. The project uses advanced technology such as drones to ensure a high degree of accuracy and transparency, and integration of drone imagery with platforms such as GIS to create a concurrent, updated and multi-layered database for all urban areas. With the support of SDAs, door-to-door surveys were conducted, and the dwelling units were numbered to integrate household data on GIS maps. Additionally, geo-fencing of current slum boundaries of about 2000 slums was undertaken using drones that captured high-resolution images of the slums, both for marking the external boundary of the slum and for detailed mapping of the slum itself.

14.7 CREATING A CULTURE OF PARTNERSHIP FOR IMPLEMENTATION OF SCHEMES

The HUDD has realised and benefitted from multiple partnerships in the implementation and scaling up of each of its programmes since 2014. The partnerships have not only been with municipalities but also with technical not-for-profit agencies and other state government departments. This has built a culture of partnership between the department and a set of other agencies who have been involved in learning through implementation alongside the department, based on their sectors of interest.

As documented throughout the chapters of this book, each programme has had the support of external partners. A key aspect has been to continuously focus on building a professional competence-based relationship, which included sharing of experiences and insights in an open manner. This method of working required the reduction of vertical hierarchy-based communication and contribution. This strategy of frank exchange through an open platform of discussion was immensely useful, given that technical resources at the HUDD are low and its field expertise limited. Over a period, through this culture of partnership in implementation, teamwork and joint ownership of the progress that was established for each programme, improved performance within the team was incentivised. This novel strategy has contributed immensely to the efficiency and efficacy in the implementation of the different programmes. It has also established the HUDD as an important partner for most of the top technical not-for-profit organisations working in India on urban development. This culture of teamwork of professionals working alongside government officers has matured over the years and has become a central organising principle for all new programmes at the HUDD.

From a time when Bill & Melinda Gates Foundation (BMGF) and Scaling City Institutions for India (SCI-FI) were working in partnership with the HUDD on one programme (in 2014), the partnerships have surged and now cover most technical agencies interested in urban development in India. These agencies include Tata Trusts, Omidyar Network, USAID, Foreign, Commonwealth & Development Office (FCDO UK), Urban Management Centre, Janaagraha, e-gov foundation, National Institute of Urban Affairs (NIUA) and several others, while BMGF and SCI-FI have also scaled up their work with the HUDD. Therefore, the HUDD and its cultural transformation has in many ways helped broaden its own work while benefitting from expertise available from multiple organisations.

The culture of partnership gained by the HUDD has led to fruitful engagement with other state government departments on joint programmes. For instance, the collaborations with the department of Panchayati Raj and drinking water have led to a formal MoU between the two departments and UNICEF and SCI-FI. Another example is the Forest Department and HUDD partnership in delisting slums on non-reserved forest lands within cities. Here too the officials of the two departments are working in close partnership to solve problems that the urban poor living in slums have been facing.

Professionalism, competence and passion are the key virtues promoted by this way of working; this makes HUDD governance processes stand out among other departments and other states. Building this culture of partnership in various tiers – from government departments and ULBs to the community level – has also led to new leadership, which have imbibed these principles, emerging at all levels.

14.8 CITIZEN FEEDBACK SYSTEM

A dedicated Feedback Cell launched on 1st August 2021, with presence in all 360 institutions which includes 114 ULBs, 237 Public Health/WATCO Section Offices and nine Development Authorities along with six call centre executives in operations. Currently, following feedbacks for following services are included:

(i) Housing (Awaas): delay in payment
(ii) Trade license: issues relating to submission of documents
(iii) Holding tax: assessment-related issues
(iv) Building plan approval (five corporations + nine development authorities (DAs))
(v) Water supply connection: pipeline issues
(vi) MUKTA: payment-related issues

The HUDD has introduced its a dedicated feedback management mechanism which aims to not only ensure the satisfaction of citizens on all urban services but also ensures quality of urban services. To date, more than 0.1 million calls have been corresponded with key stakeholders related to urban governance system; to date 1748 adverse feedback cases have been addressed.

Index

V

W

IWA Publishing's authorised EU representative for General Product Safety Regulations is Diane D'Arras, 15 rue Duret, 75116 Paris, France, e-mail: safety@iwap.co.uk.

Printed and bound by CPI Group (UK) Ltd, Croydon, CR0 4YY
12/05/2026
02108877-0001